THE ANSCHÜTZ
GYRO COMPASS.

HISTORY.

DESCRIPTION.

THEORY.

PRACTICAL USE.

The Apparatus is fully patented in all the principal Countries
of the World.

◎

ANSCHÜTZ & CO.,

Kiel, Germany.

1910.

Preface.

This Book is published to explain the principle, construction, and practical use of the Anschütz Gyro Compass.

The pages which follow are chiefly translated from the German publication by Anschütz & Co., in Kiel, but some further explanation on the subject of precession is given, and also information in greater detail on some points where questions have been asked by those interested in the apparatus.

The Chapter on Theory sets out the calculations made by Herr M. Schuler, on which the design of the Gyro Compass in its present successful form is based ; this chapter has been dealt with by Mr. Harold Crabtree, whose work on Gyrostats is well known, and by Mr. Alfred Lodge, late Professor of Pure Mathematics at Coopers Hill College. Thanks are due to these gentlemen for the time they have given to the subject, in rearranging the chapter so as to accord with English mathematical symbols and practice.

To Commander Chetwynd, R.N., Superintendent of Compasses, and Commander G. M. Marston, R.N., most grateful acknowledgement is given for numerous important suggestions on those portions of the book which deal with the practical use of the Compass on board ship.

G. K. B. Elphinstone.

Elliott Brothers,
    London, December, 1910.

Contents.

History of Gyro Experiments.

The best known experiment carried out by the great French philosopher, Foucault, is undoubtedly his employment of a pendulum for demonstrating the rotation of the earth in the year 1851. It is not so well known that he conducted a long series of researches with Gyrostats, though the work he did with these instruments was of immense importance. The great practical difficulties inseparable from Gyrostat experiments stood in the way of success of other investigators working on the same lines as Foucault, and for this reason comparatively little attention was paid to the whole subject for many years; for it must be remembered that early Gyrostats, being set in motion by winding up a string, or some similar device, could only run for short periods of time, and even then at comparatively low rates of speed.

Foucault himself undoubtedly encountered the greatest possible difficulties in his experiments, and it was chiefly due to his great powers of deduction that he was enabled to lay down theories extending far beyond the points reached experimentally.

The first law he laid down states that **any Gyrostat** possessing three degrees of freedom, or in other words, being free to move in all three planes, and **unaffected by the force of gravity, must indicate the rotation of the earth in a manner similar to that demonstrated by the pendulum** in his celebrated work; that is to say, the Gyro would continue with its plane of rotation fixed in space, while the earth turned round under the Gyrostat.

He further arrived at the conclusion that any Gyro with only two degrees of freedom, or in other words, free to move in **two planes only, will at any place on the earth's surface,** other than the two poles, **tend to set itself with its axis of rotation parallel to the axis of the earth itself, by reason of the relative rotations of the two bodies.**

It may be mentioned that the experiments carried out by Foucault in 1852 were actually suggested by Lang, of Edinburgh, in the year 1836, but were not tried practically by him.

As the use of steel in ship construction increased, so the importance of indicating **direction** by some appliance which would be unaffected by the varying magnetic influences was more and more realised, and on this account experiments were carried out by various investigators with Gyrostats having two degrees of freedom, and also three degrees of freedom, in the hope that in some such device the desired substitute for the Magnetic Compass would be found.

Amongst others Dr. Anschütz commenced experiments in the year 1900, on the lines laid down by Foucault, as regards a Gyrostat with **three degrees of freedom,** and though such an apparatus could never have been considered as a substitute for a Magnetic Compass, still it might have been of considerable service in furnishing definite fixed lines in space, which could have been used for comparing bearings or maintaining a course already definitely known ; at the same time it must be borne in mind that the construction of a Gyrostat, so that its centre of gravity and centre of suspension are absolutely coincident, presents enormous, if not insuperable, difficulties.

Appliances developed on the schemes indicated above, though capable of giving fairly satisfactory results, had a tendency to become so complicated, that Dr. Anschütz decided it was impracticable to pursue the subject on these lines.

The turning point in the development of Gyro Compass design came in the Spring of 1906 ; when Dr. Anschütz for the first time applied to a Gyro with **three degrees of freedom** a second Gyro with only **two degrees of freedom,** which had the effect of directing the whole system into the Meridian line, as foretold by Foucault's work, and the work of the years which have followed, has shown clearly that the use of such a Gyro with only **two degrees of freedom** is the correct solution of the problem.

A difficulty which had to be overcome was pointed out by Dr. Martienssen, in the "Physikalische Zeitschrift" (Jahrg 7, No. 15), that the Gyro with two degrees of freedom is affected by all other forces which are brought to bear on it by the movements of the ship as well as by the earth's rotation ; these forces setting the Gyro swinging, and therefore making its indications unreliable, and at an early stage of the experiments conducted by Dr. Anschütz this difficulty was kept in view.

To be of practical value a Gyro Compass must possess a very large gyroscopic resistance, strongly opposing any attempt to tilt its axle to an angle, and the friction of the suspension system must be as small as possible. These two facts lead to the result that if the Gyro be deflected for any reason a long way out of the Meridian line, its swinging motion to and fro will last a very long time, and while such swinging takes place many new forces may be brought to bear on the whole system.

The problem, therefore, of adapting a Gyrostat as a substitute for a Magnetic Compass appeared impossible until a successful method of "damping" the swinging motions could be applied, and in early experiments a second Gyro was employed for this purpose. Subsequently this was done away with, and at the present time the "damping" is an essential feature of the single Gyro used in the Compass.

Many further improvements followed, which were tried experimentally, and carefully tested on board ship, and eventually, in 1908, a very exhaustive series of trials, extending over four weeks, was carried out on board the German battleship *Deutschland*, since when, as the result of the success of these trials, the use of the Anschütz Gyro Compass has been rapidly extended in numerous ships of the

B

German and other Continental navies ; and, at the time that this book is printed, apparatus is under construction for the British Admiralty for use on a considerable scale.

In the German navy, the apparatus is in use for steering purposes—on the bridge, in protected positions behind armour, and also between decks.

Seldom has the application of a new principle to a practical problem come at a more opportune moment, for, up till recently, the difficulties of applying proper compensation corrections to the Magnetic Compass on warships could be overcome to a very considerable extent, but now, with the enormous increase in the size of warships, and the great masses of moving steel in use in modern guns and their shields, correct adjustment of the Magnetic Compass becomes a much more difficult problem.

In the last few years the importance of the submarine boat has increased enormously, and, owing to the essential features of their construction, as well as the great number of electric motors on these craft, an accurate means of determining direction by some " nonmagnetic " appliance is of great importance ; this is being appreciated by many of the large navies, who are installing the Gyro Compass in their submarines.

The foregoing notes may serve to explain the great interest which has been shown in the practical application of well-known laws to the Anschütz Gyro Compass. In subsequent pages are notes as to the actual use of the apparatus in practice on board ship, and a description of its mechanical and electrical details.

Gyrostats.

Principles of Precession.

It is of course well known that a rotating Gyrostat always endeavours to maintain its axle in the same direction. A familiar example of this is found in the diabolo.

For the explanation of the dynamic principles which come into action, a short popular description of the nature of **precessional motion** is given, which, while not claiming to be in any way complete, gives the practical side of this problem in a simple manner.

Precessional motion is most easily studied by means of the Gyrostatic Top illustrated below, which can be purchased anywhere.

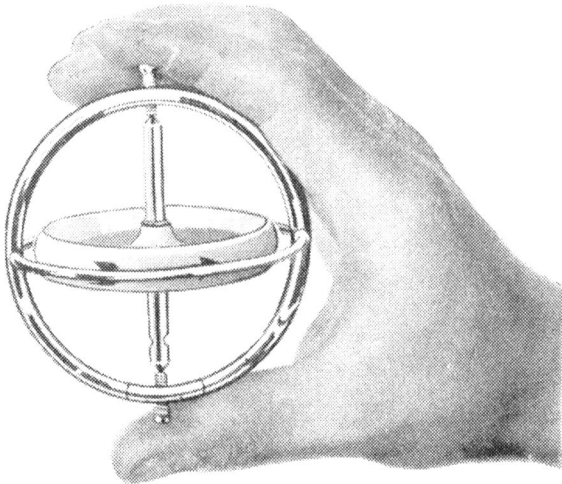

Fig. 1.
Gyrostatic Top.

If the top be held as shown in Figure 1 with the thumb under one centre screw and the first finger over the other, and a good spin be given to the wheel, it will be felt immediately that the top offers considerable resistance to any attempt to turn its axle from its original position, while it exerts a pressure as if it wanted to twist itself out of the hand. On closer observation it will be seen that the axle always tries to move at right angles to the force used to turn it, and this motion is called "precession."

It will be observed, on experiment, that, if the axis is merely moved parallel to itself, no resistance at all is offered by the Gyrostat.

In the following illustrations, Figures 2 to 6, the direction of rotation of the gyro wheel is the same in all cases.

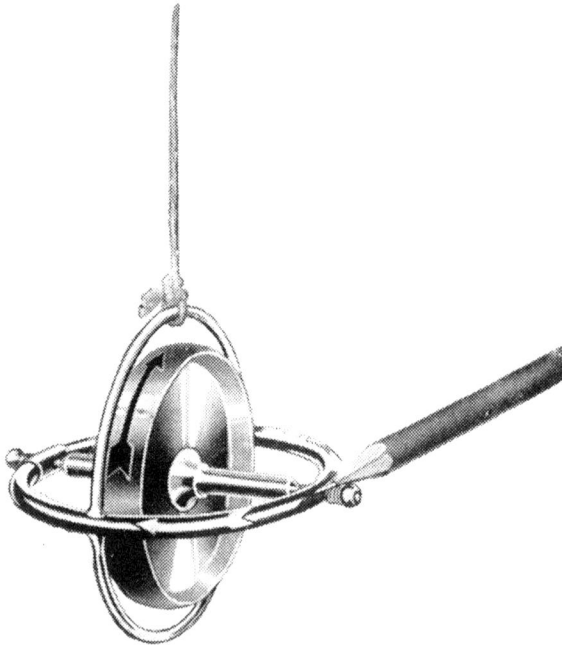

Fig. 2.

Gyrostatic Top.

With curved arrows indicating direction of rotation and direction of precession when one end of the axis is depressed by means of the pencil point.

In all the illustrations straight arrows are used to indicate the direction in which a force is applied to tilt the Gyro, while curved arrows indicate the consequent precession.

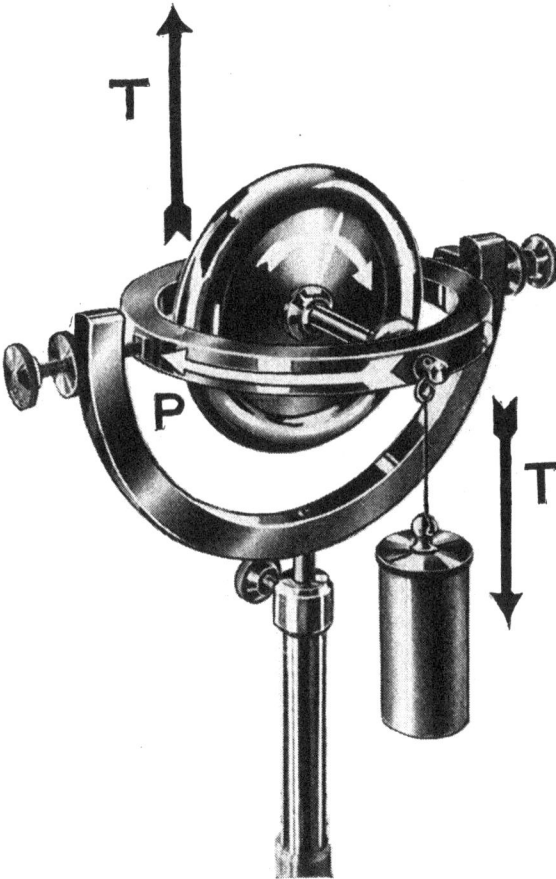

Fig. 3.

Gyrostat.

With weight suspended from one end of axis—illustrating precession, as in Figure 2.

Figures 2 and 3 illustrate the direction in which a Gyrostat precesses when an attempt is made to turn its axle from its original position. In Figure 2 the force to tilt the axle is shown as applied with a pencil, and in Figure 3 a weight is shown hanging from one end of the axle of the well-known Wheatstone Compound Gyrostat, used for lecture-room demonstrations.

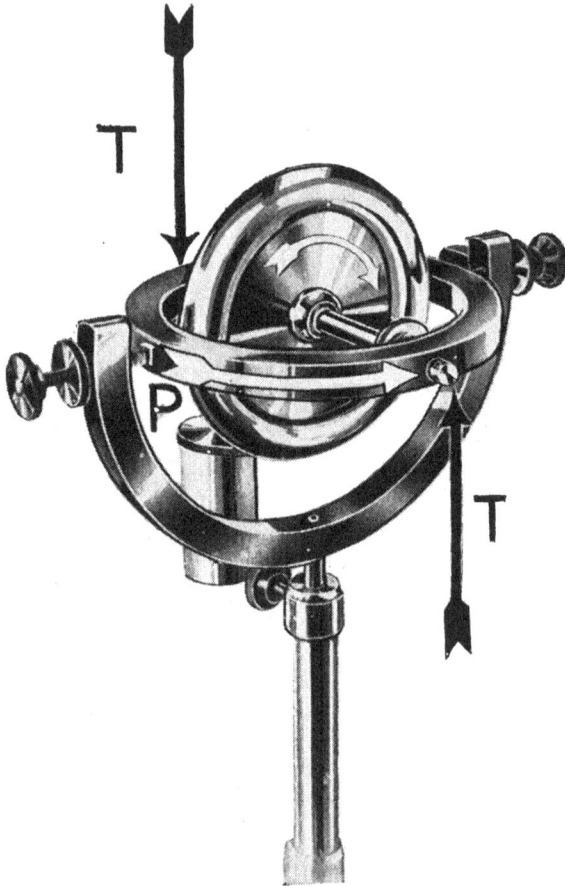

Fig. 4.

Illustrating precession in the opposite direction when pressure is applied to axis in reverse direction to that shown in Figure 3.

It must be remembered all through the consideration of the action of the Gyro Compass that precession continues all the time a force is applied to tilt the axle, and ceases as soon as the force no longer acts.

Fig. 5.

The two Figures, 5 and 6, show converse cases to those illustrated on the preceding page—where a turning force, applied as shown by the arrows T, is brought to bear on the system, causing precession in the direction of the arrows P.

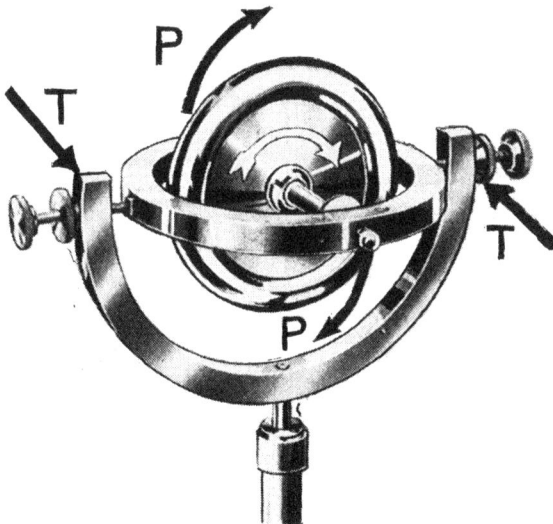

Fig. 6.

The Anschütz Gyro Compass.

Perhaps the most striking feature of the Gyro Compass, regarded as a navigational instrument, is that, unlike any previous compass, it points to the true North Pole of the earth.

The ordinary mariner's compass merely points to a certain spot known as the magnetic pole, from the direction of which the true North can be deduced. The Gyro, however, avoids this operation by indicating the true North direct.

This is obviously a great advantage, when one remembers the continual secular change in " variation," and in the other magnetic elements affecting the compass needle ; and the changes due to alteration of geographical position.

Fig. 7.

The Compass Cards are marked 0 to 360°

The general use of the Gyro Compass would enable much simplification to be made in navigational charts and sailing directions which are at present based on magnetic bearings.

General Description of the Principles of the Gyro Compass.

In order to make a description of the Gyro Compass really intelligible, the reasons for the **existence** of a **directive force** must be explained first of all.

Foucault's theory contains the general statement that " Every free rotating body, when subjected to some other or new turning force, tends to set its axis of rotation parallel to the new axis of rotation by the shortest path, so that the two rotations take place in the same direction."

It is this principle which governs the practical working of the Gyro Compass.

In the Gyro Compass, it should first be understood that the Gyro itself is carried upon a float free to move in a bowl of mercury, so designed that, as long as the Gyro is **not rotating,** the whole moving system is free to swing in every direction like a pendulum.   See Figure 11, page 25.

The centre of gravity of this whole moving system, is **below** the metacentre.

The Gyro is mounted at the lowest point of the moving system with its axle horizontal, and it therefore would swing back to this horizontal position if disturbed and then left to itself; that is to say, while the Gyro is **not rotating** the force of gravity always keeps the axle of the Gyro horizontal, and therefore the float and compass card, which are rigidly connected to it, take a corresponding position.

Reason for one end of the Axle always turning towards the North.

The illustration, Figure 8, is intended to explain general principles.

The sphere represents the earth as seen from above the North Pole, and a Gyrostat is supposed to be rotating at the equator in the direction indicated by the arrow on its periphery.

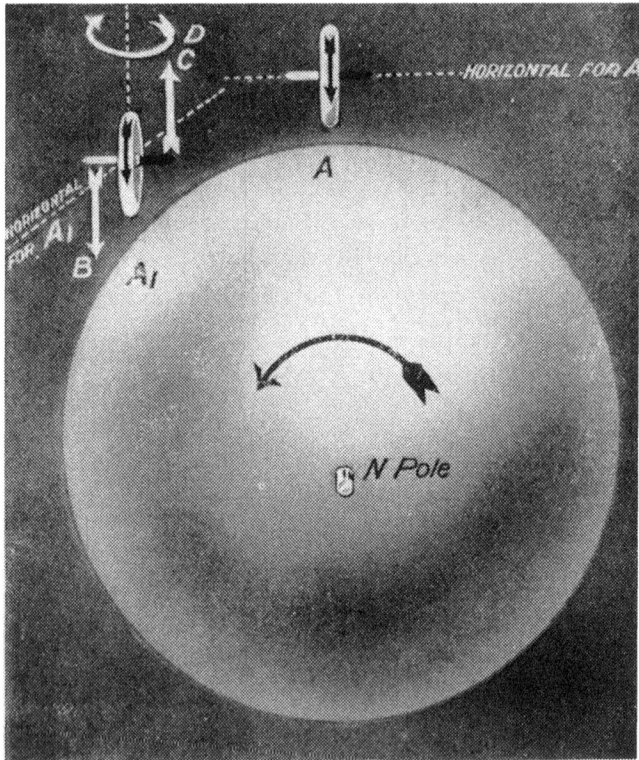

Fig. 8.

When the Gyro is at position A, its axle is horizontal and stands east and west. If, after a certain interval of time, the earth has rotated until the Gyro is at the position A₁, then, in the case of a Gyro with three degrees of freedom, *i.e.*, uniformly suspended and free to turn in all directions, the axis would no longer be horizontal as regards the surface of the earth, but the condition would be as shown in the Figure at A$_1$, the Gyro having kept its axle parallel to the original position which it occupied when at A, or in other words, the dark end of the axle would have dipped downwards from the horizontal position.

As explained above, the Gyro Compass is acted upon by the force of gravity, which tends to keep the axle horizontal ; and this action we may describe as a " couple," represented by the two straight arrows B and C ; under the influence of this the Gyro " precesses " in the direction of the curved arrow D (see notes on Precessional Motion, Figures 3 and 4, pages 13 and 14), the effect produced being exactly similar to that obtained by depressing one end of the Gyrostatic Top, Figure 2.

The direction of this precession will continue the same until a condition is arrived at when the action of gravity has no further pendulum effect on the suspended system, which clearly is when the axle of the Gyro is again horizontal. This happens in the following way :—In Figure 8 the rotation of the earth is causing the dark end to dip downwards ; when, however, the axle swings through the meridian, it will be seen that the earth's rotation causes the dark end to begin to rise, and thus the axle becomes eventually again horizontal at the same angle from the meridian on the new side as it started at originally on the old side, viz. 90°.

When the axle is horizontal, the precession, or swing to or from the meridian, comes to an end, and in all cases where this occurs (except when the axle is horizontal in the meridian plane) the action of gravity begins to tilt the axle of the Gyrostat in the reverse direction, thus reversing the direction of the precession and causing the moving system to swing back towards the meridian again.

This can be followed by imagining the Gyro in position

A_t in Figure 8 turned round, so that the white end of the axle dipped downwards from the horizontal position. Under such a condition the precession would be in a direction opposite to the arrow D. Figures 3 and 4, pages 13 and 14, may serve to make this clear.

Should the axis of rotation of the Gyro not commence by standing parallel with the equator in the starting position

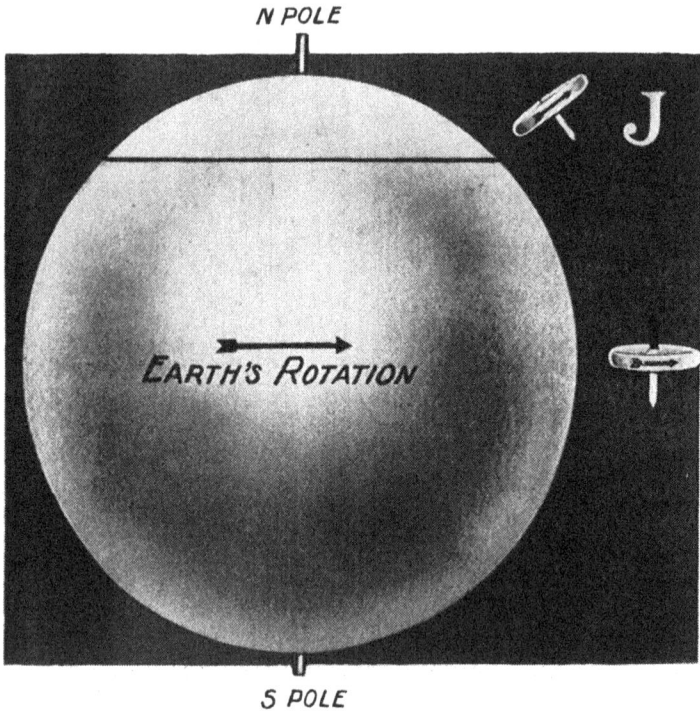

Fig. 9.

A, a little consideration will show that, no matter in what position the Gyro happens to stand, except **due north** and **south** the earth's rotation will cause a precession towards the meridian; one end of the axis of the Gyro always turning north and the other south, according to the precessional law.

The existence of such directive force can only be observed

when the speed of rotation of the Gyro is high, and all precautions for the elimination of friction are taken; under ordinary conditions, with small models, the directive force must be taken for granted, as it is too small in amount to be easily observed.

If now the Gyro in Figure 8 be considered on some parallel of latitude other than the Equator, as illustrated at J, in Figure 9, the force of gravity keeps the axis of rotation of the Gyro horizontal on the earth's surface, and turned into the meridian line by the action of the directive force described above.

It should be stated that the explanations in this chapter are only approximate; a complete analysis of the motions is given in the chapter on the Theory, see pages 51–81.

The directive force, which the rotation of the earth exercises on the Gyro, diminishes as the Poles are approached, because the higher the latitude, the smaller the actual distance moved by the Gyro (in space) in a given time. At the Poles themselves **no** directive force exists, because actually at the Poles the horizontal plane is not subjected to any angular tilting, and therefore there is no tendency for the axis of rotation of the Gyro to be tilted by the action of the earth's rotation. Also at the Poles no difference of compass direction exists, every line being a meridian.

Demonstration Model illustrating Directive Force.

For the purpose of illustrating the influence of the earth's rotation on a Gyro, the model shown in Figure 10 has been constructed. The large ring A can turn on the base B and is intended to represent a meridian line on the globe.

C is a small Gyro which has inside it a three-phase motor, D is a motor generator for converting continuous current into three-phase for the Gyro, and KK are small springs intended to represent the force of gravity of the model globe.

When the springs K **are detached** from the small ring G, the arrangement represents a Gyro with three degrees of freedom, for the Gyro can be turned in any one of three directions :—

1. Round its axis of rotation, or axle.
2. About a perpendicular axis, because the ring E can turn in the foot F.
3. About an axis at right angles to the Gyro axle, because the ring G which carries the Gyro is mounted on pivots.

If now the Gyro is set in motion by supplying its motor with electric current it will be noticed that turning the meridian ring A has only a very small effect upon the Gyro (due to the friction of the bearings), and that the Gyro axle always continues to point approximately towards the selfsame part of the room.

Now stop the rotation of the Gyro, and **attach** the **two springs** K ; then one degree of freedom is suppressed, and the Gyro is under the conditions referred to in the foregoing description, as having two degrees of freedom ; the springs themselves represent the effect of gravity, which in the Gyro Compass tends to keep the Gyro axle horizontal, as explained on page 17.

As the effect of gravity is towards the centre of the earth, so the springs on the model pull towards the centre of the meridian ring A, and keep the Gyro axis horizontal as regards the model globe (or tangential to the ring's periphery).

Now if the Gyro be set rotating and a slow turning motion imparted by hand to the ring A, we shall see at once that the Gyro, under these conditions, after a few swings to and fro, sets its rotation axis parallel to the meridian ring of our model.

Until the Ring A **is** turned, **no** tendency exists for the axis to set itself round.

23

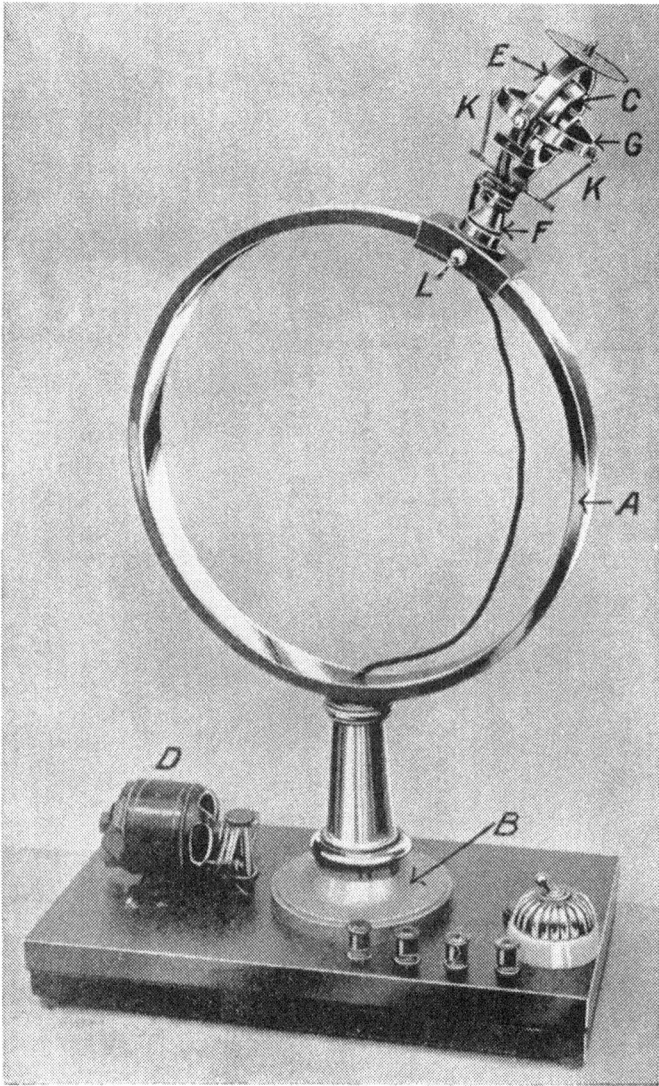

Fig. 10.

The foot F can be adjusted to any position round the ring A and clamped there by screw L, so that the Gyro can be placed at any desired "latitude," and the diminution of the directive force can be observed when the Poles are approached.

The design of the model does not admit very readily of reversing the spin of the wheel. The effect of the latter can, however, be obtained by turning the meridian ring in the reverse direction. If this be done the Gyro swings through an angle of 180°; the end which was pointing north now pointing south. Thus the Gyro may be said definitely to have a North and a South Pole at opposite ends of its axle, though one must bear in mind that it is a " Pole " due to rapid rotation and not to any magnetic effect.

Description of the Practical Construction of the Gyro Compass.

As soon as it is understood that the axis of rotation of a revolving Gyro, when correctly suspended, possesses a directive force due to the earth's rotation, in some respects analagous to the magnetic needle, the various points in the diagram (Figure 11) can easily be followed. This diagram shows a vertical section through the centre of the Gyro Compass as constructed for use on board ship.

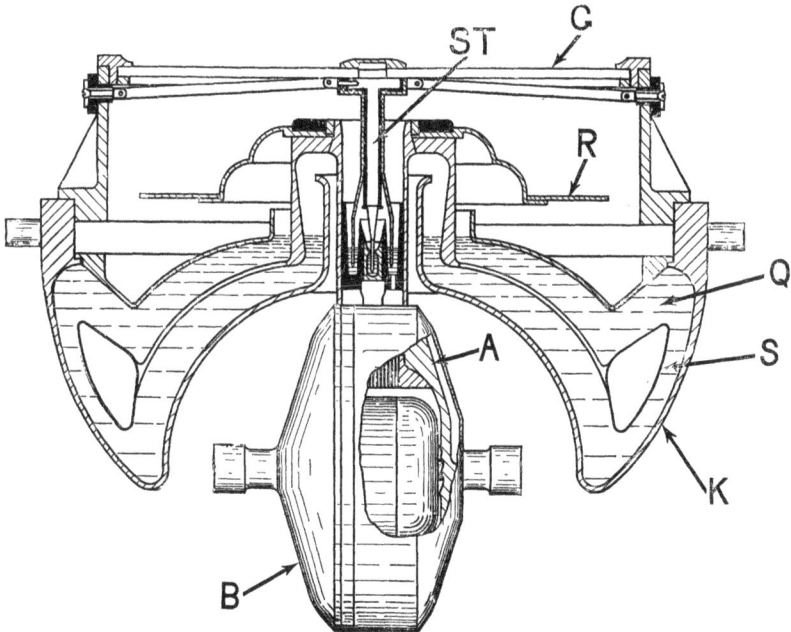

Fig. 11.

In Figure 11 the case B carrying the bearings of the Gyro A is supported by means of a stalk below the float S, which consists of a circular hollow steel ring attached to a dome shaped upper part. This floats in the circular bowl K also made of steel and filled with mercury as at Q.

C

Rigidly attached to the float S and the Gyro A is the compass card R, which therefore follows all the movements of **float and Gyro**; the axle of the Gyro is directly under the **north** and **south** line on the **card**.

Through the glass G on the top of the instrument, the divisions round the compass card R can be seen, and also the lubber line; a small spirit level mounted on the card makes it possible for any tilting of the axis of rotation of the Gyro from the horizontal to be observed.

The level is not shown in Figure 11, but can be seen by reference to Fig. 37, page 84, at S.

The whole mercury bowl K is carried on gymbals in the well-known manner, and the outer gymbal ring is borne by springs from the binnacle case, so that it is to a great extent protected from damage due to violent shocks; as far as the gymbals and methods of suspension are concerned, these do not differ materially from the design of any ordinary Magnetic Compass, and one might almost describe the Gyro Compass as a "Liquid Compass" in which the magnetic needle is replaced by a revolving Gyro with its axle always pointing true north and south.

In order to keep the whole floating system central a steel stem ST is fixed centrally in the top glass G, and the lower end of the stem dips into a small mercury cup carried on the top of the float. A similar connection is effected by a steel tube mounted concentrically with the stem ST and a second mercury cup. These two sets of connections are electrically insulated from one another and from the general metal portions of the apparatus.

These two connections carry two phases of a three-phase current to the motor of the Gyro; the third phase reaches the motor through the mercury bowl, mercury and float. On this account, the whole instrument is insulated from the binnacle, first by insulation at the gymbal supports, and, as a further safeguard, by insulating supports for the suspension springs where these are secured to the binnacle.

The motor of the Gyro is not shown in Figure 11, but consists of a very small three-phase motor, the stator of which carries the windings, and is mounted inside the case B, so that all the connections can be rigidly made. The rotor is rigidly fixed into the inside of the Gyro flywheel itself.

The speed of rotation of the Gyro should be about 20,000 revolutions per minute. It is constructed, spindle and all, from one solid piece of special nickel steel, so that there is no chance of anything working loose. The axle is provided with ball bearings made of a specially hard steel, so as to withstand wear for long periods of time ; allowance is made for expansion due to heat, and means are provided for replacing all parts of the bearings if required.

The axle of the Gyro is of the de Laval type, or a "flexible axis," so that the centre of gravity of the whole rotating mass coincides with the rotation axis as soon as a certain critical speed is exceeded. Even though the axle is relatively weak, the Gyro, while running, is not sensitive to shocks, because while even the very shortest possible shock lasts the Gyro has made several revolutions (333 per second) and therefore any bending tendencies neutralise one another.

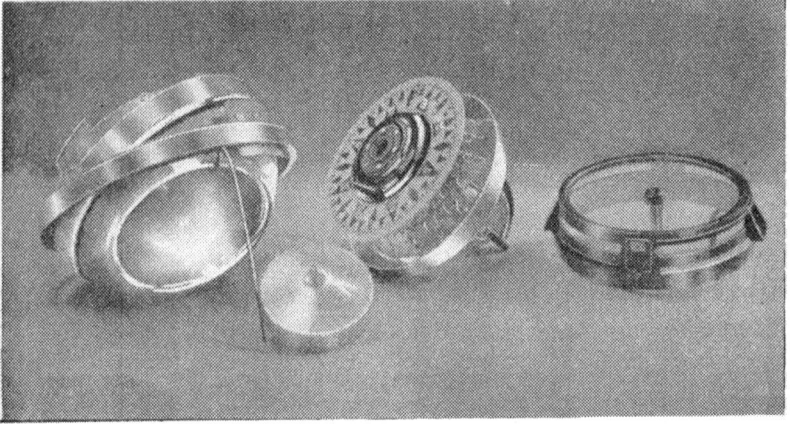

Fig. 12.

Figures 12 and 13 illustrate the chief parts of the Gyro
Compass and, taking these from left to right—

    Mercury bowl with gymbal rings.

    Floating system, consisting of Gyro in its casing, float
        and compass card.

    The Gyro itself without its casing.

    The top cover with the central stem.

Fig. 13.

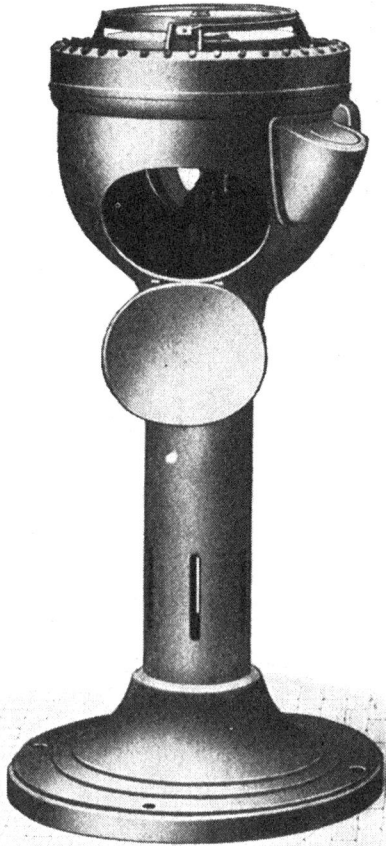

Fig. 14.

Figure 14 shows a Gyro Compass in a binnacle with the cover removed, and a door opened to give access to the interior.

Fig. 15.

Figure 15 illustrates the general arrangement of the binnacle containing the master compass or transmitter, together with the reversible motor, commutator and transmitting gear.

Description of Damping Device.

The friction between the suspended moving system and the mercury in the bowl is not sufficient to cause a visible decrease in the amplitude of the oscillations of the Gyro on either side of the meridian, and a condition would continue indefinitely as illustrated by the curve, Figure 16, if no artificial damping were employed.

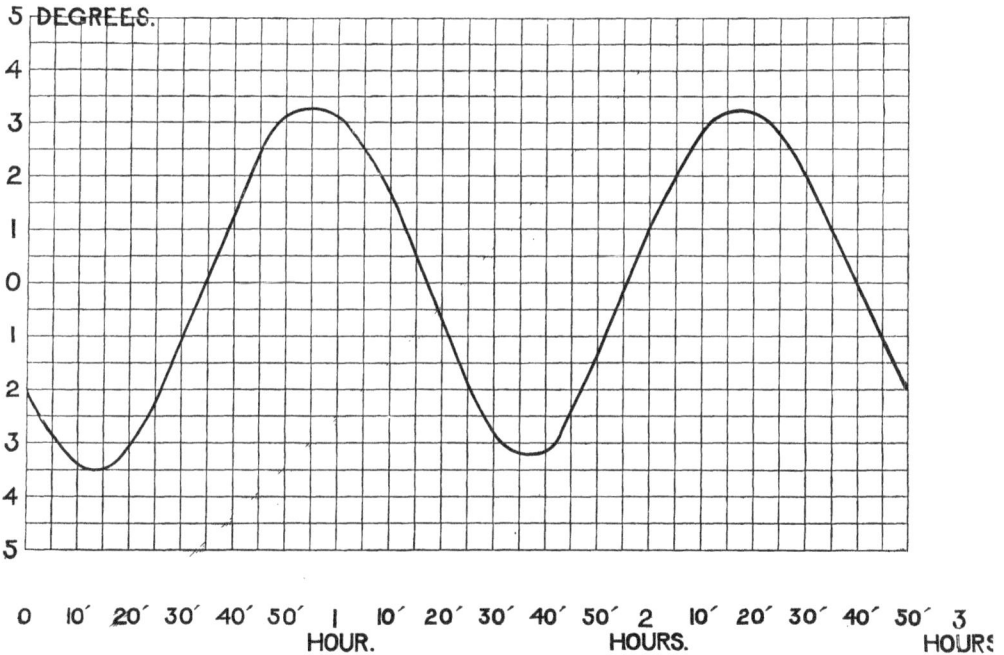

Fig. 16.

Undamped Oscillations.

An artificial damping is applied as shortly described below. In the chapter on Theory (pages 51–81) a more complete explanation is given.

The effect of this damping is shown in the curve, Figure 17, which illustrates a Gyro Compass settling down to its correct reading in three hours from the time of starting up the Gyro motor, the compass at the moment of starting pointing nearly 45° away from the true meridian.

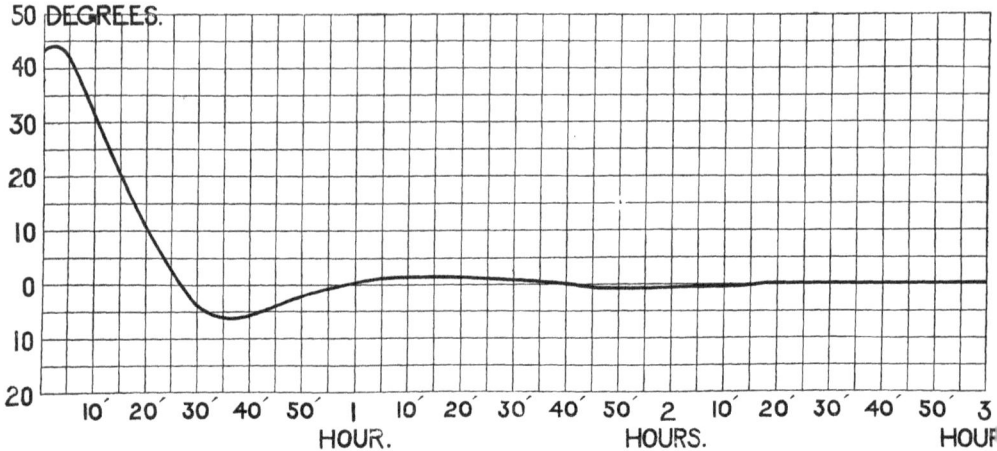

Fig. 17.

Damped Oscillations.

Cut near the centre in the sides of the case p are two holes, g, for the admission of air (see Figure 18) and on the periphery of the case an outlet is provided. The Gyro acts as a high speed centrifugal blower, and a strong air blast is set in motion, incidentally serving to keep the motor of the Gyro cool, and at the same time made use of to **damp** the swinging of the Gyro on either side of the meridian.

A constant stream of air issues from the opening in c, and this opening is divided into two parts, a and b, by the plate u, carried by the pendulum arm d.

The arm d hangs from accurately constructed bearings, very free from friction, and the arm is so balanced that, when the axle of the Gyro is horizontal, the two openings, a and b, are equal, and the stream of air divides itself equally between them, one part on each side of a vertical centre line through the whole moving system.

Should the axle of the Gyro not be horizontal, which it will be recollected is the case when the axle is precessing to or from the meridian, the small pendulum d swings to one side and the plate u automatically enlarges one of the openings a b, and closes the other; the two streams of air are then no longer equal ; the difference of their reaction forms a turning couple round the vertical axis of the system.

Fig. 18.

a, b	Variable outlets for air blast.	f, r	Gyro bearings.
c	Outlet pipe.	g	Inlet opening for air.
d	Pendulum arm.	q	Terminal of Gyro motor.
e, s	Oil cups for Gyro bearings.	p	Gyro case.
		o	Mercury bowl.

Under the influence of this couple a motion is set up opposing the precession, in such a direction as to bring the axle of the Gyro once more horizontal. In this manner the oscillations of the Gyro to either side of the meridian are powerfully "damped," as illustrated in the curve given above, Figure 17.

In the Gyro Compasses where a large binnacle is used. as in the case of the master compass used in the transmission system, the damping arrangement is somewhat simpler, the pendulum d is done away with, and the outlet for the air is a small rectangular opening in the Gyro case; the effect of the two arrangements is exactly the same, provided sufficient space exists in the binnacle for the air blast to be free from effects due to air currents in the casing.

Explanation of Necessary Corrections.

The following short description is given of the reasons for certain corrections being necessary. In the chapter on Theory the mathematical treatment of these is set out very fully.

Latitude Correction.

A small correction is necessary in changing from one latitude to another, as will be seen from the following considerations, analysis shows that this need not be taken into account for a change of latitude of less than 10°.

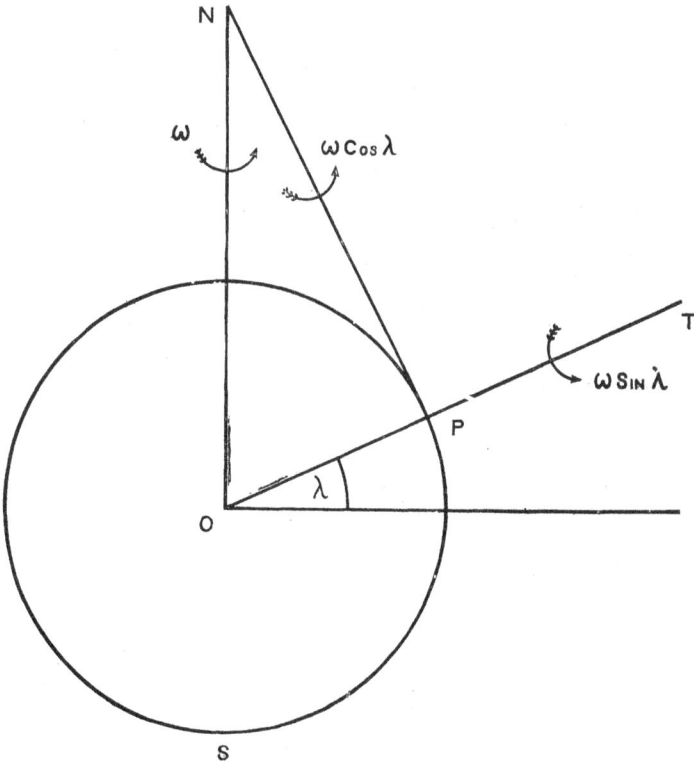

Fig. 19.

Figure 19 shows that the angular velocity ω of the earth about its axis can be resolved at any place P, into

$\omega \sin \lambda$ about the vertical O P T.

$\omega \cos \lambda$,,   ,, meridian line P N.

Since the earth is always turning from west to east underneath the compass, which tends to maintain its direction in space, it follows that the axle will be **left behind** on the **east** side of the meridian, unless a precessional velocity, $\omega \sin \lambda$, can be imparted to it by some couple. The moment the axle begins to lag behind the meridian a portion of $\omega \cos \lambda$ causes a tilt which introduces the gravity couple, **but simultaneously** the damping couple tends to destroy this by making the axle horizontal. Hence there will be some position of lagging behind the meridian, for which the damping just maintains the axle at that tilt at which the gravity couple can cause the necessary precession about the vertical. By the action of the earth's rotation the gravity couple is always tending to be increased by an amount which the damping is simultaneously tending to diminish, and thus the couple is kept constant to cause the constant precession $\omega \sin \lambda$.

It is of course clear that this precession is only constant for a given latitude. By placing a small weight on the Gyro in the binnacle as at t, Figure 18, a couple could be introduced to suit any particular latitude, and yet the axle can remain horizontal ; but any change of latitude from that for which the weight is adjusted will necessitate a small correction.

The table for correction of latitude is given below :

Latitude 60° north   ·6 (36') easterly.
  ,,   50°   ,,
  ,,   40°   ,,   ·5 (30') westerly.
  ,,   20°   ,,   1°·1 (1° 6')   ,,
  ,,   0°   ,,   1°·6 (1° 36')   ,,
  ,,   20° south 2°·1 (2° 6')   ,,
  ,,   40°   ,,   2°·7 (2° 42')   ,,
  ,,   60°   ,,   3°·8 (3° 48')   ,,

All Gyro Compasses are adjusted for a latitude of 50° north.

In the foregoing pages the Gyro Compass has been considered as being at rest on the earth's surface.

When, however, the compass is mounted on board ship, and the ship moves over the earth's surface, certain corrections have to be applied to its readings.

Ship's Movements at Uniform Speed.

(Angle δ.)

If the axle of the Gyro is pointing north and south it is clear that any movement of the ship due east or west is merged in the movement due to the earth's rotation, and is negligible in amount.

A movement, however, due to steering a northerly course, will produce a couple deflecting the Gyro Compass westwards, owing to the fact that the axle always tends to keep to its direction in space. Similarly in the case of a southerly course, the compass is deflected eastwards.

This deflection depends simply on the latitude, speed, and course of the ship, and in no way on the design of an individual instrument; it can therefore be calculated for all cases, and tables for the necessary correction are given on pages 88–92.

In the case say of a north-east course, it is only the northern component of the speed which causes the deflection from the true course; this deflection is, for a given latitude and speed, proportional to the cosine of the angle between the ship's course and the north.

For example, in latitude 10° (north or south), with a speed of 12 knots, the angle δ on a northerly course is ·8°, and on a north-east course $\cdot8 \times \cos. 45° = \cdot8 \times \dfrac{1}{\sqrt{2}} = \cdot5$.

Change of Ship's Speed.

(Ballistic deflection.)

From the foregoing it is clear that changes of ship's speed, either due east or west, will not affect the compass, as the pendulum motion of the suspended system is then about an axis parallel to the axle of the Gyro, and therefore the Gyro axle moves parallel to itself, but northerly or southerly changes of speed must theoretically have an effect on the readings.

Considering the case of a ship steaming north and stopping suddenly, the suspended system of the Gyro has a tendency to swing forward from its own inertia, and the direction of the resulting precision is such as to cause a westerly error.

This is called the " ballistic deflection," and depends entirely on the design of the instrument. In practice the damping method described above makes it possible for the ballistic deflection to coincide with the " Angle δ " for the new speed, so that the correction for " Angle δ " can be applied immediately, and does all that is necessary.

Description of Transmission Instruments.

The introduction of a rapidly rotating Gyro in place of the Magnetic Compass allows of the compass being acted upon by a far greater directive force than is possible with a magnetic needle. In point of fact, the directive force on the Gyro Compass is some **fifteen times** as great as in the case of a good Magnetic Liquid Compass in some position free from all disturbances from surrounding iron.

The Gyro Compass is completely free from all magnetic influences, but powerfully resists any alteration of direction of its axle, this fact applying equally to movements in the horizontal and also in the **vertical** plane; in other words, the Compass Card can only swing vertically up and down with the north and south line as an axis; therefore the east and west points on the card can move up and down, while the north and south points on the card always maintain the same horizontal position; and on this account a contact point on the east and west line on the card can only move straight up and down and not round in a circle. This phenomenon renders simple the construction of a system of transmission mechanism which would be impossible with a Magnetic Compass where the card can roll about in any direction, with the point of support as a centre.

The apparatus used in the Gyro Compass consists of a transmitter, attached to the Master Compass, and receivers, which can be connected electrically as shown in the connection diagram at the end of this pamphlet, Figure 40, page 104. This arrangement allows the Master Compass and Transmitter to be placed in some well-protected position low down in a ship where the use of a Magnetic Compass would be impossible owing to the influence of surrounding iron. The indications of the Master are accurately repeated on the receivers, which can be placed in various positions in the ship, such as for instance the steering position, conning tower, etc., etc.

In some large ships two complete Master Compasses and Transmitters are being installed in two different positions, with a complete duplicate system of receivers.

The Master Compass and Transmitter differs from the ordinary single Gyro Compass by the fact that the mercury bowl can be rotated without any work being thrown on the Gyro Compass, the design being such that it always turns in the same direction as the Gyro, or in other words "follows" its movement, if one may so express it for simplicity; in point of fact, it is the case or binnacle which moves, while the Gyro stands still.

The rotation is done by a reversible motor controlled by contacts which are operated by the Gyro itself, so that, if the contact on the right hand side (as seen from the centre) presses against the contact carried by the Gyro on the Compass Card, the mercury bowl is moved to the right, and the right hand contact moved with it far enough to open the

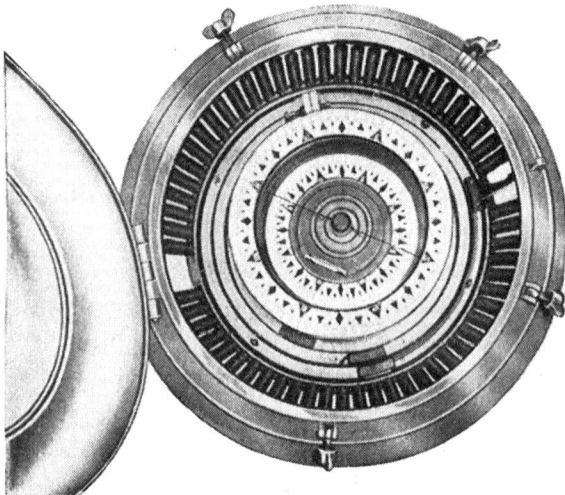

Fig. 20.

circuit, when the movement ceases, and vice versa—the diagram of connections at the end of this pamphlet illustrates this. Figure 41, page 105.

The reversible motor turns the mercury bowl at a speed which must be quicker than the rate of turning of any large ship, and therefore no **lag** takes place.

A commutator is mounted on the axle of the reversible motor which distributes currents to the mechanism of the receiving instruments so that these always turn in synchronism with the transmitter, a special design of electrical receiver being used on account of the high rate of speed at which the signals are transmitted.

The receiving mechanism is connected by means of gearing to a Compass Card. By the above scheme the indications of the Master Compass, or Transmitter, can be observed in as many positions as desired.

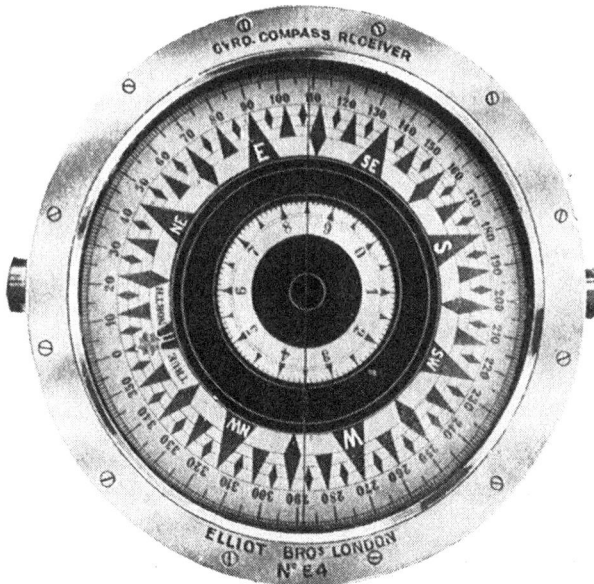

Fig. 21.

General Appearance of Receiver with Central Compass Card.

Figure 20 illustrates the general appearance of the Transmitter, and the Receiver is illustrated at Figures 21 and 22.

It will be seen in Figure 21 that a second Compass Card is provided in the centre of the dial. This makes one complete revolution for 10° alteration of course, and is divided so that an alteration of course of a few minutes is at once apparent. The employment of this " fine adjustment," so to speak, is of the greatest possible assistance in steering, and facilitates the immediate observation, and correction of every small amounts of Yaw.

When under weigh, the small central Compass Card is constantly " on the move " to and fro on account of the ship continually departing from an absolutely straight course.

So long as the movement is " to and fro," this serves as an indication that everything is working—while a continued movement in either direction shows that an alteration of course has taken place.

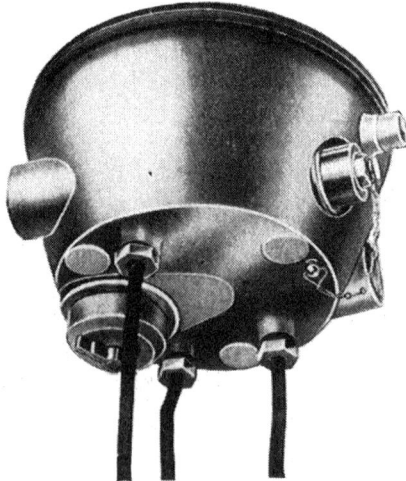

Fig. 22.

Receiver Case.

The two following illustrations make the use of the Central Compass Dial quite clear.

Fig. 23.

Illustrating the Compass Cards when the ship is on a course of 107°, the last figure 7 for a single degree on the outer card being reproduced on the Central Compass Card on a large scale.

Fig. 24.

Illustrating the Compass Cards when the ship has yawed off her course 1½°, the reading being now 108½.

The receivers are independent of position, and can, if necessary, be connected up by means of flexible cable. In the case of the Steering Compass, the dial can conveniently be inclined at an angle, or fixed vertically. The outside ring of the receiver case can be divided off to facilitate the taking of bearings, with an ordinary Azimuth mirror.

It is unnecessary to instal the receivers with the lubber point fore and aft in the ship, except in the case of an instrument used for taking bearings, which must of course be mounted centrally as in the case of a magnetic compass. In practice, however, it is usually found more convenient to instal the receivers with the lubber point fore and aft.

The Compass Cards are made of opal glass, so that electric lamps can be placed inside the case of the receivers, to illuminate the dial ; a regulating switch is provided so that any desired amount of illumination can be employed, or the light turned off if not required.

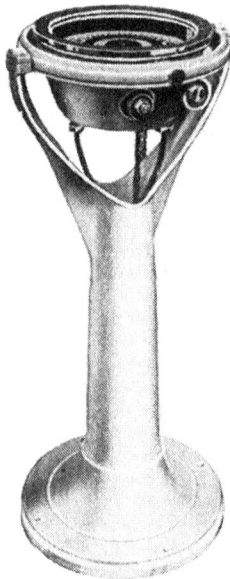

Fig. 25.

Receiver on Pillar.

General Description of Accessory Apparatus.

The three-phase current employed to run the motor of the Gyro, and the reversible motor where transmission gear is installed, is supplied by a specially designed motor generator, illustrated at Figure 26, the three-phase current being furnished at 120 volts pressure, and at a periodicity of 333~ per second.

Fig. 26.

The continuous current, or motor portion, can run off the ship's lighting circuit; and the power required for the whole installation is about 700 watts. The three-phase circuit is entirely insulated from the continuous current circuit.

The three-phase portion of the motor generator has **16 poles,** it runs at a normal speed of 2,500 revolutions per minute, the speed being shown on an indicator attached to its shaft—as the motor in the Gyro has only **two poles,** it follows that the speed of this latter is $16 \times 2,500 = 20,000$ revolutions per minute.

2

Each equipment comprises, besides the motor generator, a switchboard, with a starting switch, and regulating rheostat; one arrangement of this switchboard is shown at Figure 27.

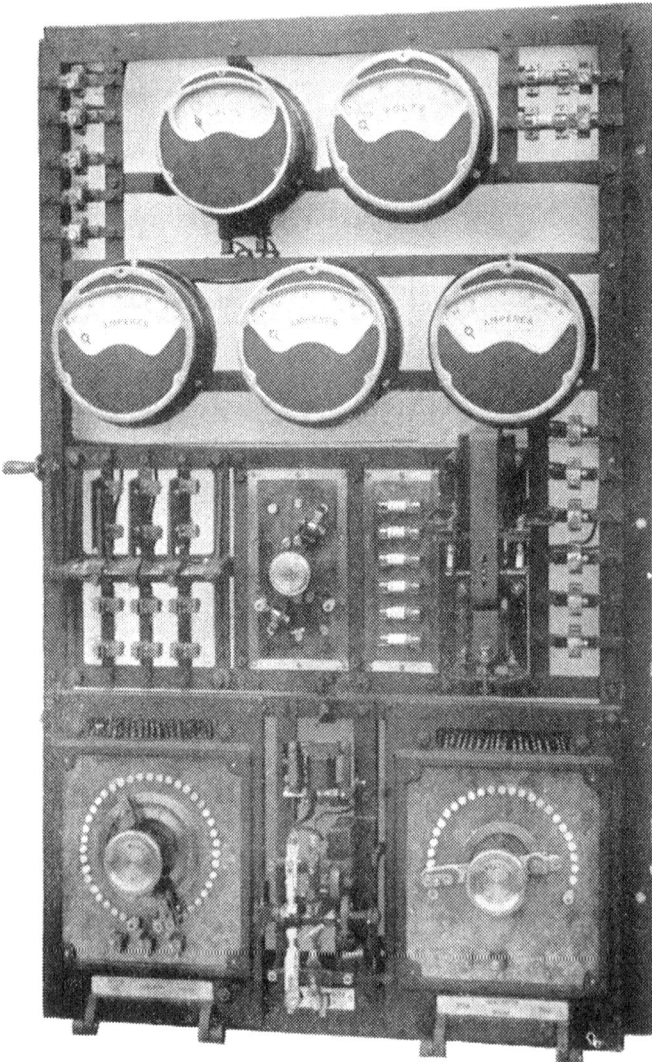

Fig. 27.

In the arrangement illustrated, 3 separate amperemeters

are arranged, one in each phase—a voltmeter is provided with a switch so that the pressure between any two line wires can be read. Fuses are provided, in duplicate, with a switch so that the circuit can be instantaneously changed over should one fuse fail. The burnt out fuse can then be changed while the circuit is maintained through the other set.

Junction and Fuse Box.

A junction and fuse box mounted near the binnacle contains fuses in each of the circuits leading from the commutator of the transmitting device to the individual receivers. There is a fuse in **each** circuit so that in case any circuit fails, it is disconnected and the other circuits remain available for use.

Further details of this are given in the chapter of the maintenance of the apparatus, see pages 98, 103, and diagrams of connections are given at Figures 40, 41, 42, 43, 44, 45, pages 104–109.

Constructional Details.

To enable the Gyro Compass to give accurate results in practice, very special attention has been given to every detail of construction, the very highest possible accuracy being necessary in the manufacture of all parts. Many of these parts are of intricate form and have to be specially constructed with a minimum of weight.

Considering the Gyro itself, it is obvious, on account of the high speed (20,000 r.p.m.) at which this revolves, that very special conditions arise, and enormous centrifugal forces have to be dealt with; the stress to which the rim is subjected amounting to some 10 tons per square inch.

A very special steel is employed to ensure a sufficiently high factor of safety, to ensure that sufficient margin exists, and it may be of interest to state that an experimental Gyro was run up far above its normal speed, until it eventually gave way.

To obtain this experimental result a great many special arrangements had to be made. The Gyro was run in a vacuum, as otherwise, on account of the air friction increasing with the square of the speed, too much power would have been required, and, further, a special motor generator had to be built to give the enormously increased periodicity necessary for the enormous rate of speed.

Figure 28 illustrates the Gyro at the end of the experiments. The case twisted, but not torn apart. The case checked the rotation of the Gyro as soon as it touched. It was found that some **five times** the normal power had to be applied; this in itself is a proof that no yielding of the material employed can take place in practice, as the motor generator installed cannot give sufficient output to cause so high a speed.

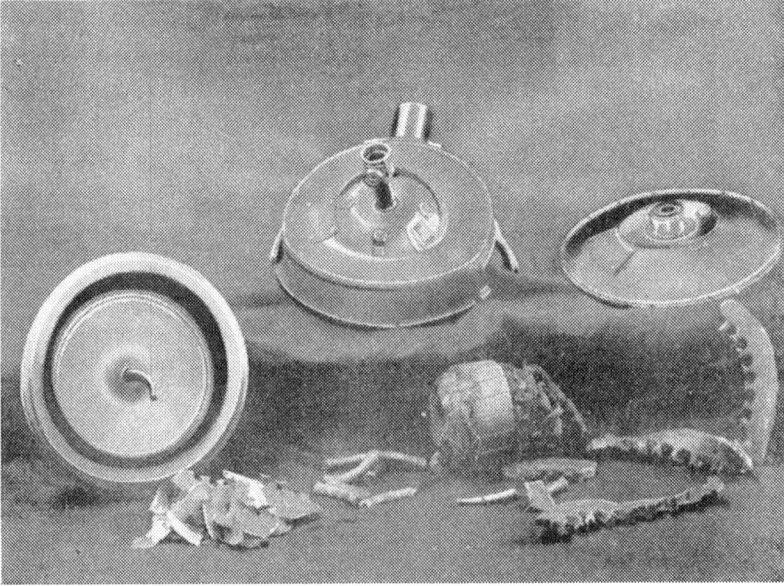

Fig. 28.

The peripheral speed of the Gyro, at its normal rate, is 500 feet per second, or 340 miles per hour ; the air friction is so great that 95 per cent. of the work done by the Gyro Motor is absorbed in this manner, and a curious result follows—that the surface of a Gyro which has run a few thousand hours is noticeably smoother than when it left the finishing process in the grinding machine before being put into use. There is no doubt that this polishing is due to the actual air friction on the steel itself.

The ball bearings, upon which the shaft of the Gyro runs, require the most minute attention in their construction ; the balls have to be gauged to almost inconceivably small limits, and special precision appliances have been devised to examine the spherical condition of the balls.

The Gyro Motor itself calls for electrical work of quite a special order. No motor was available commercially which could run 20,000 revolutions per minute, and to get such a machine into the very small space available without undue heating presented a very difficult problem. No idea can be formed when the finished Gyro is seen of the immense number of experiments which were necessary to bring its practical details to a satisfactory state of perfection. Amongst other points, the usually accepted text-book figures for the constants of the iron in the motor no longer held good with a periodicity as high as 333~.

Theory.

Equations of Motion of the Gyro Compass.

This Chapter is intended for those readers who are already familiar with the general mathematical treatment of rotating bodies.

Elementary explanations are to be found in the following works :—

Spinning Tops—Professor J. Perry (Romance of Science Series).

Dynamics of Rotation—Professor A. M. Worthington (Longmans, Green & Co.).

Spinning Tops and Gyroscopic Motion—Harold Crabtree (Longmans, Green & Co.).

Über die Theorie des Kreisels—F. Klein and A. Sommerfeld (Leipsic).

A Table of Symbols employed in the following investigation and a complete Schedule of the Equations used are given on pages 77–81.

From what has already been described on page 18, it is clear that, damping being forces neglected, if the Gyro be spinning " right handed " looking from the centre towards one end, and is placed pointing west, then the end in question will turn to the north, dipping **downwards** towards the earth ; while, if it be pointing east, it will turn to the north inclined **upwards** from the earth : see Figure 29, which is drawn for this latter case, when the Gyro axle has passed through the meridian from the east to the west side.

In Figure 29, M represents the metacentre and G the centre of gravity of the movable parts of the system, so that the distance M G is the metacentric height, denoted by a.

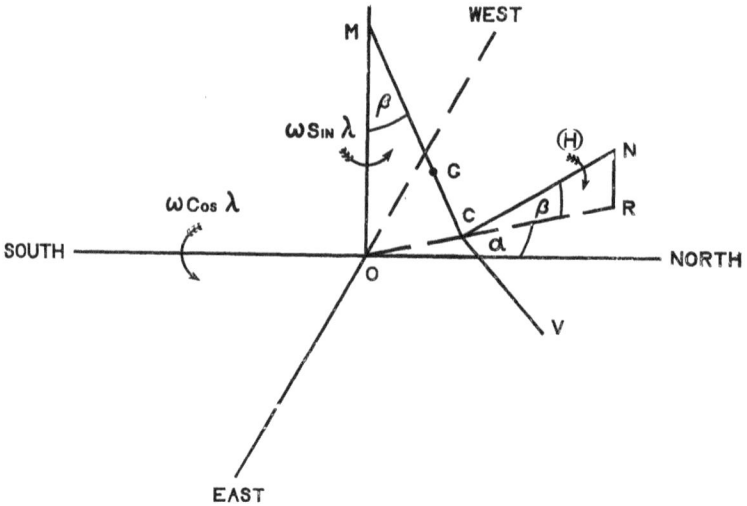

Fig. 29.

The axle of the Gyro is represented by the line C N, inclined at an angle β to the horizontal, and M G, C N are drawn in the same vertical plane, which makes an angle a with the meridian, on the west side of it. In this position a will be considered positive. The angle M C N is a right angle.

When the plane M C N is vertical, as in the Figure, then, if we neglect friction and damping effects, which are discussed later, page 57, the only couple acting on the system, so as to cause precession, is—

$$M \ g \ a \ \sin \beta$$

where M is the mass of the movable parts and g is the acceleration due to gravity.

If the plane of M C N is not vertical, but is slightly tilted about a line through M parallel to C N, no further precessional effect takes place, since the axle of the Gyro is merely moved parallel to itself. The arm of the couple M g a sin β will be **slightly** altered, but in the following investigation the amount of this alteration is negligible, so that the only couple causing precession is

$$M \ g \ a \ \sin \beta$$

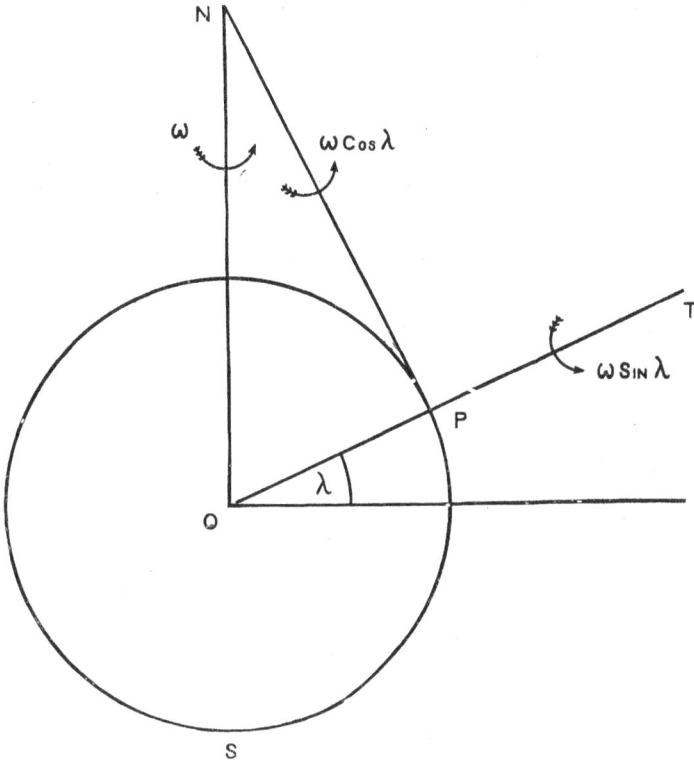

Fig. 30.

Components of the Earth's Rotation.

Figure 30 represents the rotation (ω) of the earth resolved into two components, about the meridian and about the vertical, at any place P whose angle of latitude is λ.

Further, the component $\omega \cos \lambda$ can be resolved (see Figure 29) into—

(i) $\omega \cos \lambda \sin a$ about C V the horizontal line perpendicular to C R.

(ii) $\omega \cos \lambda \cos a$ about C R.

The component $\omega \cos \lambda \sin a$ represents the rate at which the horizontal plane is dipping, or rising, relatively to the axle of the Gyro ; while the component $\omega \cos \lambda \cos a$ only affects the numerical value of the spin of the Gyro relative to the earth, and is negligible.

The velocity $\omega \sin \lambda$ is the rate at which the meridian is turning underneath the Gyro Compass, so that when the latter is damped it tends continually to lag behind the meridian on the east side.

It thus appears that if the axle is to keep pace with the meridian (apart from deflections from its position of rest relative to the earth) some couple must be brought into action to cause a precessional velocity in space equal to $\omega \sin \lambda$. This is discussed on page 60.

Equations of Motion.

The correct speed for the rotation of the wheel in the Anschütz Gyro Compass is 20,000 revolutions per minute, giving an angular momentum about the axle C N (Figure 29), which is always 100,000 times, and sometimes 1,000,000 times, as great as the component of the angular momentum due to precession. We may therefore, in writing down the equations of motion, consider that the angular momentum of every moving part is a negligible quantity compared with that of the Gyro wheel itself about its axle, which angular momentum we will denote by H.

It is a well known theorem[*] that : If a body which has angular momentum I ω, or H, about an axis O X, be under the action of a torque K about a perpendicular axis O Y, then the **angular momentum will be rotated** about the third perpendicular axis O Z with angular velocity Ω determined by the equation.

$$K = H \Omega$$

---

[*] A full discussion of this method of dealing with a rotating body is given in Crabtree's "Spinning Tops and Gyroscopic Motion," page 37, sqq. In the same work the theory of gyroscopic resistance is employed (page 114) to deduce the general equations of motion, of which the equations in this article are a particular case.

O being, either a fixed point in the body, or its centre of gravity. The sign of Ω is positive when, H and K being drawn in the same sense, the former sets itself towards the latter.

It must be remembered that the above is only true of **steady** motion, when the precession has **already** been started ; in other words :—to **maintain** the precession Ω of the angular momentum H, a couple K is required such that K = H Ω.

The expression H Ω is frequently called the Gyroscopic resistance which H offers to being turned about the axis of K.

If the motion is **not** steady, the couple K must be equated to all gyroscopic resistances + the rate of change of angular momentum about its own axis ; but, as already stated, in our problem all angular momenta are negligible in comparison with H, the angular momentum of the Gyro Wheel.

We are now in a position to write down the equations of motion for the Gyro Compass.

Referring to Figure 29, and considering the angular momentum of the system about C R as being rotated round the vertical line through C by the external couple due to gravity, we have :—

$$\text{Angular momentum rotated} = H \cos \beta$$

$$\text{Velocity of precession (in space)} \quad \ldots \quad \ldots \quad = \frac{d\,a}{d\,t} + \omega \sin \lambda$$

$$\text{Rotating couple} \quad \ldots \quad = M\,g\,a \sin \beta$$

$$\text{Hence } H \cos \beta \left(\frac{d\,a}{d\,t} + \omega \sin \lambda \right) = M\,g\,a \sin \beta$$

or since, in practice, owing to the subsequent damping, β is a small quantity—

$$(1) \qquad H \left(\frac{d\,a}{d\,t} + \omega \sin \lambda \right) = M\,g\,a\,\beta$$

The left-hand side of this equation is the gyroscopic resistance offered by the axle to the external couple, which would otherwise cause it to dip. This illustrates the fundamental equation

$$\text{External Couple} = \text{Gyroscopic Resistance.}$$

Again, we may consider the angular momentum about C N (Figure 29) as precessing round the horizontal line C V, perpendicular to C N, under the action of a zero couple about M C. We then have—

Angular momentum rotated = H

Velocity of precession (in space) $= \dfrac{d\,\beta}{d\,t} + \omega \cos \lambda \sin a$

Rotating couple ... ... = zero (neglecting all friction).

Hence

(2) $\quad H \left(\dfrac{d\,\beta}{d\,t} + \omega \cos \lambda \sin a \right) = 0$

This equation shows that, when friction is neglected, the velocity of rise or dip is merely the rate at which the earth is turning from or up to the axle; the axle itself is not rising or falling in space.

Differentiating (1) we have—

(3)
$$H \frac{d^2 a}{d\,t^2} = M\,g\,a \frac{d\,\beta}{d\,t}$$

and eliminating $\dfrac{d\,\beta}{d\,t}$ from (2) and (3)

we obtain

(4)
$$\frac{H^2}{M\,g\,a} \cdot \frac{d^2\,a}{d\,t^2} + H\,\omega \cos \lambda \sin a = 0$$

The expression H has not been cancelled, in order to keep the dimensions of the equation in proper form. It will be seen that the dimensions of $\dfrac{H^2}{M\,g\,a}$ are those of moment of inertia, while $H\,\omega \cos \lambda \sin a$ is a couple, and represents the gyroscopic resistance offered to turning the axle about M C (Figure 29) away from the meridian. It will therefore be called the **righting moment** of the Gyro Compass.

The form of equation (4) shows that the motion corresponds to the swing of a pendulum through the position of equilibrium $a = 0$, i.e., through the meridian, the moment of inertia of the pendulum being $\dfrac{H^2}{M\,g\,a}$, which will be denoted by I.

The righting moment H ω cos λ sin a is a maximum when $a = 90°$; *i.e.*, when the axle points east and west, and in that case it is equal to H ω cos λ. It corresponds to the righting force of the Magnetic Compass, and will be called the **righting co-efficient** of the Gyro Compass, denoted by R. The actual righting moment at any instant is R sin a.

In practice a is never very large, consequently if T_o be the time of a complete oscillation of the Gyro axle, given by the above equation,

$$
\begin{aligned}
T_o &= 2\,\pi\,\sqrt{\dfrac{I}{R}} \\[2mm]
&= 2\,\pi\,\sqrt{\dfrac{H}{M\,g\,a\,\omega\,\cos\lambda}}
\end{aligned}
$$

(5)

Damping of the Oscillations.

Our investigations so far apply to undamped oscillations, as we have not considered friction at all. The natural fluid friction of the mercury in the Gyro Compass is too small to cause any appreciable diminution in the amplitude of oscillations, as described page 31, and illustrated in Figure 16. In consequence of this, an artificial damping is applied, as described on pages 31–34, the effect of which on the oscillations is shown by the curve in Figure 17, page 32.

It will be seen, on referring to the description of the damping, that immediately the axle is deflected from the meridian from any cause it begins to rise or dip owing to the earth's rotation, and therefore the restoring couple due to gravity is brought into play, while **simultaneously** the air blast couple tends to restore the axle to a horizontal position, diminishing the gravity couple, and so checking the swing through the meridian, thus bringing the Gyro Compass to rest by rapidly decreasing oscillations.

This position of equilibrium relative to the meridian will be called the **resting position** of the Compass.

E

From the construction of the damping arrangement, it is clear that the moment of the reaction is proportional to sin β.

The couple due to the air blasts may therefore be represented by D sin β, where D would be its value if the formula held for all values of β up to 90°.

Since β is small, the couple reduces to D β, and moreover, since the precessional speed of swing of the Gyro Compass, in space, in the azimuthal (a) plane, is from the equation (1) **also** proportional to β, it follows that the turning couple of the air blast is proportional to this precessional speed.

Since the air couple acts about a vertical axis, equation (1) is not affected by it, but equation (2) becomes changed. Hence the two equations are now,

$$(6) \quad \begin{cases} (i) \quad H\left(\frac{d\,a}{d\,t} + \omega\sin\lambda\right) = M\,g\,a\,\beta \\ (ii) \quad H\left(\frac{d\,\beta}{d\,t} + \omega\cos\lambda\sin a\right) = -D\,\beta \end{cases}$$

and by eliminating β the equation of azimuthal swing becomes

$$(6a) \quad \frac{H^2}{Mga}\cdot\frac{d^2 a}{d\,t^2} + H\omega\cos\lambda\sin a + \frac{HD}{Mga}\left(\frac{d\,a}{d\,t} + \omega\sin\lambda\right) = 0$$

It thus appears that the first two terms of equation (4) are unaltered ; that is, the swinging mass and the original righting moment are in no way affected by the damping.

The third term corresponds to the damping forces of friction on a pendulum. Its co-efficient $\frac{HD}{Mga}$ will be denoted by k.*

The third term shows also that the turning moment of the air couple on the swinging Gyro Compass is composed of two parts: **one** proportional to the rate of deflection relative to the meridian, and the **other** proportional to the angular velocity of the meridian itself **in space.**

These results might have been expected, since β (to which the air couple $D\beta$ is proportional) depends upon the swing of the Gyro Compass **in space.**

The **second** component is always present, whether the Compass swings, and whatever the speed of swing, relative to the meridian, and depends only upon the latitude and the magnitude of the damping co-efficient.

This turning moment $\dfrac{H\,D\,\omega\,\sin\,\lambda}{M\,g\,a}$ causes the resting position of the Gyro Compass to be displaced to a position inclined at an angle a_0 to the meridian, sufficient to enable this permanent part of the air couple to be balanced by the righting moment (see page 34).

The equation giving this condition is

$$
(7)\quad
\begin{cases}
H\,\omega\,\cos\,\lambda\,\sin\,a_0 = -\dfrac{H\,D}{M\,g\,a}\,\omega\,\sin\,\lambda \\[2ex]
\text{or since } a_0 \text{ is always small,} \\[2ex]
\qquad\qquad a_0 = -\dfrac{D}{M\,g\,a}\,\tan\,\lambda
\end{cases}
$$

If β_0 is the value of β when $\dfrac{d\,a}{d\,t} = 0$, we have from equation (6.i):

$$
(8)\qquad \beta_0 = \frac{H\,\omega\,\sin\,\lambda}{M\,g\,a}
$$

It follows that when $\beta = \beta_0$, a is a maximum, positive or negative; *i.e.*, the Gyro Compass is at the extremity of a swing through the meridian.

When the Gyro axle finally settles down in the **resting position,** and only in this case, a_0 is the value of a corresponding to β_0.

The value β_o which enables the Gyro Compass to stay in its resting position, can be obtained from the elementary consideration that, since the meridian is turning westwards underneath the Gyro with an angular velocity of $\omega \sin \lambda$, a precession of $\omega \sin \lambda$ in space must be imparted to the axle of the Gyro by the gravity couple if the Compass is ever to be at rest relative to the earth.

This necessitates $M \, g \, a \, \sin \beta_o \ = \ H \, \omega \, \sin \lambda$

$$\text{or } \beta_o \ = \ \frac{H \, \omega \, \sin \lambda}{M \, g \, a}$$

Only at the Equator is the resting position of the axle exactly in the horizontal plane and pointing true north, unless corrected for a given latitude, as explained below.

By adding a small weight to one side of the case containing the Gyro wheel, we can introduce a turning couple apart from the gravity effect on the Gyro itself, and thus keep the Gyro Compass pointing true north, and still with its axle horizontal; so for any desired latitude a_o can be made to disappear.

Unless, however, the amount or position of this weight is altered, a change of latitude introduces a_o, and necessitates a correction.

A Table of Corrections for navigable latitudes is given.

If the Gyro Compass is adjusted for the Equator, the errors are :

Latitude.	Error in Degrees.
60° N.	2°·2 Easterly.
40° N.	1°·1 Easterly.
20° N.	·5 Easterly.
0° Equator.	0°
20° S.	·5 Westerly.
40° S.	1° 1 Westerly.
60° S.	2°·2 Westerly.

If the Gyro Compass is adjusted for latitude 50° North :

60° N.	·6 Easterly.
50° N.	0°
40° N.	·5 Westerly.
20° N.	1°·1 Westerly.
0° Equator.	1°·6 Westerly.
20° S.	2°·1 Westerly.
40° S.	2°·7 Westerly.
60° S.	3°·8 Westerly.

It will be observed that the amount of the correction necessary for the Gyro Compass due to the change of latitude is very small, and need only be taken into account for changes of 10°. Clearly, changes of longitude cannot affect the Gyro Compass.

Oscillations of the Angle *a*.

If the angle *a* is small we get the usual differential equations for small damped pendulum oscillations.

For equation (6a) is

$$\frac{H^2}{M\,g\,a} \cdot \frac{d^2\,a}{d\,t^2} + H\,\omega\,\cos\lambda\,(\sin a - \sin a_o) + \frac{H\,D}{M\,g\,a} \cdot \frac{d\,a}{d\,t} = 0$$

and, writing a_1 for $a - a_o$, reduces when *a* is small to

$$(9) \qquad I\,\frac{d^2\,a_1}{d\,t^2} + k\,\frac{d\,a_1}{d\,t} + R\,a_1 = 0$$

employing the abbreviations mentioned above.

The solution of this equation is known. If the time is reckoned from the moment the axle passes through the resting position $(a = a_o)$ we get, when the motion is oscillatory :—

$$(10) \quad a_1 = a - a_o = A\,e^{-\frac{k\,t}{2\,I}}\,\sin\sqrt{\frac{R}{I} - \frac{k^2}{4\,I^2}} \cdot t$$

where A is a constant of amplitude.

It follows, then, that the time of a complete oscillation of the damped Gyro Compass is given by :—

$$(11) \qquad T_1 = \frac{4\,\pi\,I}{\sqrt{4\,I\,R - k^2}}$$

and the previous equation (10) can be written :—

(11a) $$a = \Lambda e^{-\frac{k t}{2 I}} \sin \frac{2 \pi t}{T_1} + a_o$$

If in equation (11) $k^2 \geqq 4 I R$, there is no real time of oscillation, but simply a non-periodic stopping; that is to say, the adjustment is " dead-beat."

This can be arrived at with the design of the apparatus already described, but is avoided for ordinary latitudes.

The reason why the latitude plays a part in this equation is that the righting co-efficient R varies as the cosine of the angle of latitude.

If $k^2 < 4 I R$, then real oscillations occur, which is the only case we need consider. See Figure 17, page 32.

These oscillations are asymptotic in character, never ceasing (theoretically), but after a short time they become so small as to be negligible.

Their amplitudes diminish in a constant ratio.

For, writing $p = \dfrac{k}{2 I}$ in equation (11a), we have

$$a = A e^{-p t} \sin \frac{2 \pi t}{T_1} + a_c$$

whence

$$\frac{d a}{d t} = - p A e^{-p t} \sin \frac{2 \pi t}{T_1} + \frac{2 \pi}{T_1} A e^{-p t} \cos \frac{2 \pi t}{T_1}$$

therefore when $\dfrac{d a}{d t} = 0$, that is to say, at the extremity of any swing, we get—

$$\tan \frac{2 \pi t}{T_1} = \frac{2 \pi}{p T_1} = \frac{4 \pi I}{k T_1}$$

If we call one of these angles η, it follows that the series of corresponding values of t is given by

$$\frac{2 \pi t}{T_1} = \eta + n \pi$$

and if t_1, t_2 be times at which any two consecutive amplitudes, a_1 and a_2, occur (Figure 31), then, the deflections being in opposite directions,

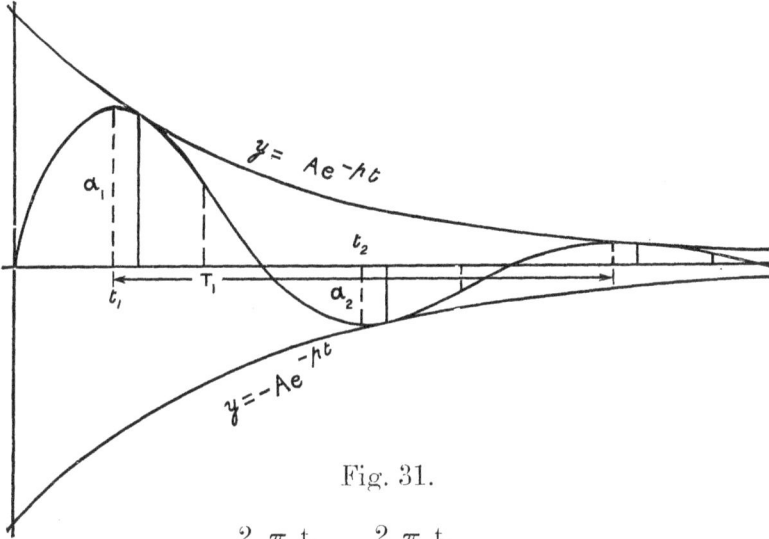

$$y = A e^{-\hbar t}$$

$$y = -A e^{-\hbar t}$$

Fig. 31.

$$\frac{2\,\pi\,t_2}{T_1} - \frac{2\,\pi\,t_1}{T_1} = \pi$$

or $t_2 - t_1 = \tfrac{1}{2}\,T_1$

and the numerical value of the ratio $a_2 : a_1$ is

$$\frac{a_2}{a_1} = e^{-p(t_2 - t_1)} = e^{-\tfrac{1}{2} p\,T_1} = e^{-\frac{k\,T_1}{4\,I}}$$

which is constant.

This constant ratio in the design of different instruments varies from 0·1 to 0·5. The natural logarithm of this ratio is in the theory of oscillations called the logarithmic decrement, and its value is in this case — ϵ, where

$$\epsilon = \frac{k\,T_1}{4\,I} = \frac{D\,T_1}{4\,H}$$

recollecting that $k = \dfrac{H\,D}{M\,g\,a}$ and $I = \dfrac{H^2}{M\,g\,a}$.

The curves $y^2 = \pm A\,e^{-p\,t}$ are boundaries of the oscillations.

A is a constant giving the amplitude of these curves when the time is zero ; and the time after which the amplitude of these curves becomes negligible depends upon A and $\frac{k}{2\,I}$

The number of intermediate oscillations varies with the latitude since it depends upon T_1 (see equation 11).

It is remarkable how the theoretically calculated curves of damping agree with the results obtained in practice.

Figure 16, page 31 shows the curve of an undamped apparatus, and Figure 17, page 32 a sketch curve of a damped instrument.

Simultaneous Equations of a and β.

Hitherto we have followed only the oscillations of the Gyro axle in the horizontal (a) plane ; the swings in the vertical (β) plane are simultaneous with these.

When there is no damping the equations are

$$(12) \quad \begin{cases} \text{(i.)} \quad a = A \sin \frac{2\,\pi\,t}{T_0} \\ \text{(ii.)} \quad \beta = A \frac{2\,\pi}{T_0} \frac{H}{M\,g\,a} \cos \frac{2\,\pi\,t}{T_0} + \beta_0 \end{cases}$$

as can be seen by putting $k = 0$ in equation (11a), and by substitution in equation (3).

We have previously determined the value of

$$\beta_0 = \frac{H\,\omega\,\sin\,\lambda}{M\,g\,a}$$

If now we set out the values of a and β in a system of rectangular co-ordinates, we obtain a series of ellipses with a constant axis-ratio. The major axis of each is at a distance β_0 from the axis of a. For every amplitude constant A there is a definite ellipse. If the Gyro does not swing we must have $a = 0$, $\beta = \beta_0$. See Figure 32, in which the origin is moved to this point.

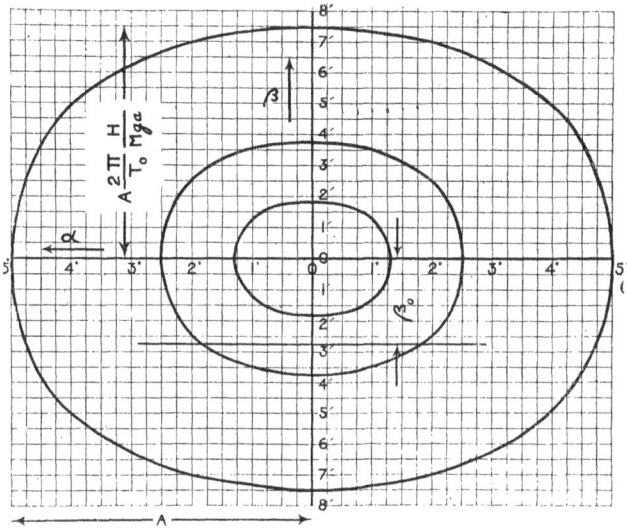

In the Figure, the Scale of β is magnified 30 times.

Fig. 32.

Similar results are obtained by taking the damping into consideration, with the difference that the end of the swinging axis describes a sort of elliptical spiral instead of a simple ellipse.

The relation between a and β in this latter case is depicted in Figure 33, and the corresponding equations are given on page 67. Referring to the figure, it will be noticed that the numerical values of a and β are reckoned from a_0 and β_0, so that they are really the values of $a - a_0$ and $\beta - \beta_0$. The figure very clearly shows that $\frac{d\,a}{d\,t} = 0$ whenever $\beta = \beta_0$, and that in the final position $a = a_0$ and $\beta = \beta_0$. The curve is not actually continued to the final position because an infinite number of convolutions are required to reach it. The character of the oscillations is also well exemplified in Figure 17, page 32.

66

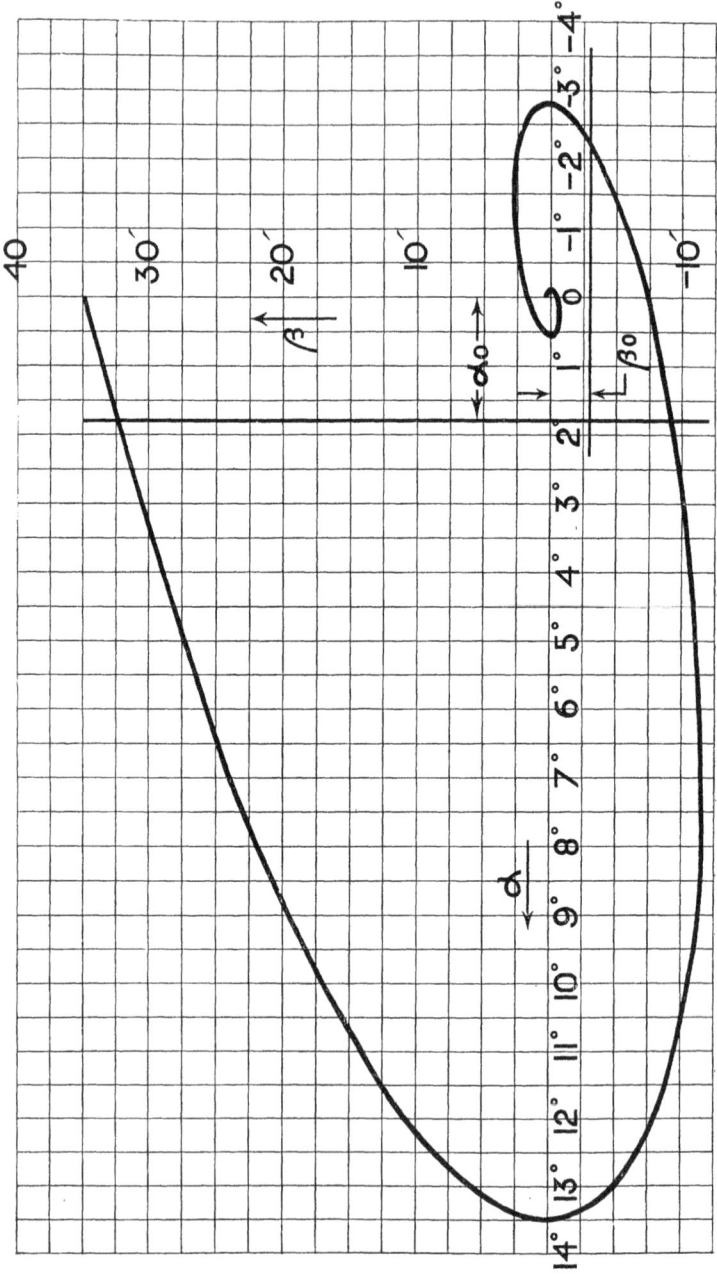

Fig. 33.

When damping is considered, the two simultaneous equations are (11a), and a combination of (6 i.) with (8), giving—

(13)
$$\text{(i)} \quad a = A e^{-\frac{k\,t}{2\,I}} \sin \frac{2\,\pi\,t}{T_1} + a_0$$

$$\text{(ii)} \quad \beta = \frac{H}{M\,g\,a} \cdot \frac{d\,a}{d\,t} + \beta_0$$

whence

$$\beta = \frac{H}{M\,g\,a} A e^{-\frac{k\,t}{2\,I}} \left(\frac{2\,\pi}{T_1} \cos \frac{2\,\pi\,t}{T_1} - \frac{k}{2\,I} \sin \frac{2\,\pi\,t}{T_1} \right) + \beta_0$$

To simplify this last equation, we have from (11) and (5):

$$T_1{}^2 = \frac{4\,\pi^2\,I^2}{I\,R - \frac{1}{4}\,k^2} \quad \text{and} \quad T_0{}^2 = \frac{4\,\pi^2\,I}{R}$$

therefore
$$\frac{4\,\pi^2}{T_1{}^2} = \frac{R}{I} - \frac{k^2}{4\,I^2}$$

that is to say,

(14)
$$\left(\frac{2\,\pi}{T_1} \right)^2 + \left(\frac{k}{2\,I} \right)^2 = \left(\frac{2\,\pi}{T_0} \right)^2$$

If, therefore, we introduce an angle ξ, such that, referring to Figure 34,

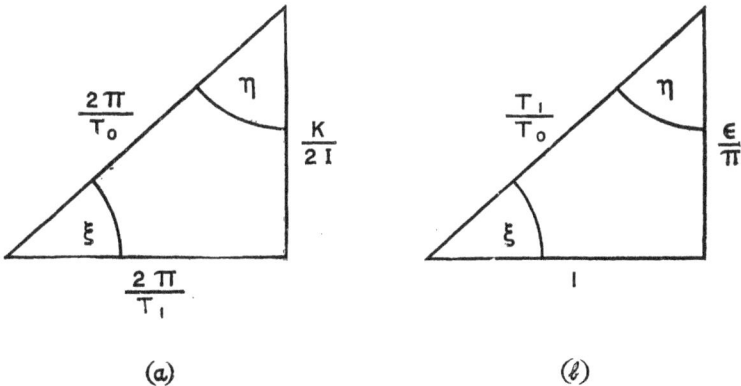

Fig. 34.

$$(15) \begin{cases} \dfrac{2\pi}{T_o} \cos \xi = \dfrac{2\pi}{T_1} \\[2mm] \dfrac{2\pi}{T_o} \sin \xi = \dfrac{k}{2\,I} \end{cases}$$

and therefore $\tan \xi = \dfrac{k}{2\,I} \cdot \dfrac{T_1}{2\pi} = \dfrac{\epsilon}{\pi}$

we can write the equations for a and β as follows :—

$$(16) \begin{cases} \text{(i)} \quad a = A\,e^{-\frac{k\,t}{2\,I}} \sin \dfrac{2\pi t}{T_1} + a_o \\[4mm] \text{(ii)} \quad \beta = \dfrac{A\,H}{M\,g\,a} \cdot \dfrac{2\pi}{T_o}\,e^{-\frac{k\,t}{2\,I}} \cos \left(\dfrac{2\pi t}{T_1} + \xi \right) + \beta_o \end{cases}$$

It should be noted that by comparing the value of $\tan \xi$ as given by Figure 34 (a) with the value of $\tan \eta$ as given on page 62, viz., $\tan \eta = \dfrac{4\pi I}{k\,T_1}$, it is clear that ξ and η are complementary angles. Hence η is inserted in Figures 34 (a) and 34 (b).

Figure 34 (b) is the same as Figure 34 (a) on a different scale and shows that

$$\left(\dfrac{T_1}{T_o} \right)^2 = 1 + \left(\dfrac{\epsilon}{\pi} \right)^2$$

The graphic solutions of equations (12) and (16) are given in Figures 32 and 33, pages 65 and 66.

The data for Figure 33 are taken from an actual Gyro Compass, and since the important magnitudes for which the curves are drawn have also a numerical interest, they are given here briefly :—

Righting co-efficient at equator $= R = 20{,}190$ dyn. cms.
Moment of inertia ... ... $= I = 404 \times 10^7$ gr. cm^2.
Period of damped oscillation $= T_1 = 4{,}110$ seconds.
Period of undamped oscillation $= T_o = 3{,}680$,,
Logarithmic decrement of
    damping given by ... $\epsilon = 1\cdot56$

Additional Corrections. Angle δ.

On pages 35–38 a simple explanation has been given of the nature of the necessary corrections, the manner in which these are applied, and Tables of Values are given in the chapter on Practical Use on Board Ship, pages 89–92.

Since the righting moment of the Gyro Compass is dependent on the earth's rotation, it follows that the motion of a ship on her course must influence the indications of the instrument. Every motion over the earth's surface can be resolved into two components, one of which lies in the N—S direction, and the other in the E—W direction.

The latter component only adds to, or subtracts from, the earth's velocity, and hence increases or decreases the righting moment by a **very small amount** which is negligible.

On the other hand, the N—S component is equivalent to an angular rotation round an E—W axis, and must be added vectorially to the angular speed of the earth.

Hence the position in which the righting moment on the Gyro Compass will be zero is now, not the meridian, but a horizontal line inclined at some angle δ to the meridian. If v is the component of the ship's speed in the N—S direction, and E the earth's radius, then $\dfrac{v}{E}$ is the corresponding angular velocity of the ship about the E—W axis; and the angle δ between the position of zero righting moment and the meridian is, since δ is small, given by

$$(17) \quad \delta = \frac{v}{E} \cdot \frac{1}{\omega \cos \lambda} \quad \text{as illustrated in Figure 35.}$$

The angle attains only a small value even with the highest obtainable speed, but where really accurate compass readings are required it must be taken into account, as explained in the chapter on Practical Use on Board Ship.

The value of the angle δ is in no way dependent on any particular Gyro Compass, but is only a geometrical relation between the ship's speed and the earth's rotation. From

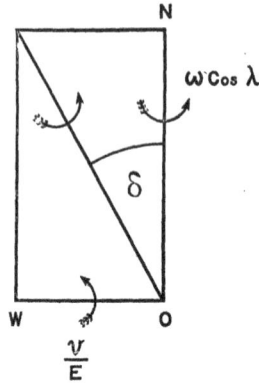

Fig. 35.

Figure 35 it is clear that δ is westerly for a northerly course and easterly for a southerly course.

Ballistic Deflection.

Apart from the angle δ dependent on speed and course, there is the ballistic deflection μ caused by a **change** of speed of the ship. The two deflections are quite distinct, but the Gyro Compass is designed so that in practice only the angle δ has to be considered.

All acceleration or retardation effects act at the point of suspension of the moving system of the Gyro, while the forces of inertia act at the centre of gravity of the moving system, which, it will be remembered, is at a distance a below the point of suspension or metacentre.

Now, if the Gyro Compass points steadily due north, an acceleration in the E—W direction merely causes the moving system to heel about an N—S axis, without causing any alteration of the direction of the axle of the Gyro.

If, however, the acceleration is in the N—S direction, *i.e.*, in the direction of the Gyro axle, then the inertia of the moving system causes a precessional motion to take place owing to the tilting couple on the Gyro axle.

If we call this acceleration γ, we have by the laws of precession :—

(18) $$H \frac{d\,a}{d\,t} = M\,a\,\gamma$$

therefore if μ is the corresponding change in a,

(19) $$\mu = \frac{M\,a}{H} \int_0^T \gamma\,d\,t$$

The righting moment and damping are not here considered, so that the equation only holds exactly for the limiting case when the time T of the disturbance can be reckoned as infinitely small compared with the time of oscillation of the instrument.

The integral $\int_0^T \gamma\,d\,t$, in the limit when $T = 0$, measures the velocity generated by the impulse. The equation then may be written

(20) $$\mu = \frac{M\,a}{H} (v_2 - v_1)$$

where v_1, v_2 are the velocities in the N—S direction before and after the impulse.

The angle μ, as may be seen from the equation, is not dependent on the earth's rotation, or the latitude, as in the case of angle δ, but only on the design of the apparatus, and on this account it can be reduced to a definite magnitude.

To arrive at a convenient formula, we insert the time of undamped oscillations as given in equation (5), viz. :

$$T_0 = 2\,\pi\,\sqrt{\frac{H}{M\,g\,a\,\omega\,\cos\lambda}}$$

Whence

$$\mu = \frac{4\,\pi^2}{g\,\omega\,\cos\lambda} \cdot \frac{1}{T_0^2} (v_2 - v_1)$$

or, if we take T_o as the time of undamped oscillations of the particular Gyro Compass at a definite latitude (λ_o), such as 50° N, then

(21) $$\mu = \frac{4\,\pi^2}{g\,\omega\,\cos\lambda_o} \cdot \frac{1}{T_o^2}\,(v_2 - v_1)$$

thus emphasizing the fact that μ does not vary with the latitude.

The formula above shows that μ is inversely proportional to the square of this standard value of T_o.

In the series of curves given below, Figure 36, the relation between the values of μ and different values of T_o taken as standard are shown for four different values of γ.

Since, now, the maximum possible speed of the ship and consequently the greatest value of $(v_2 - v_1)$ are known, we can see, by equation (21) above, how far the Gyro Compass can be deflected under the most unfavourable conditions of altering course, stopping, getting under way, &c., &c. The value of μ, which corresponds to the maximum possible value of $v_2 - v_1$, can never be exceeded, and by the choice of a sufficiently long period T_o of swing, it can be reduced to any desired magnitude.

As soon as the ship's speed v_2 is reached, the Gyro swings in accordance with the curve shown in Figure 33, page 66, into its new resting position, which is then displaced from the position a_o by the angle δ, corresponding to v_2.

Now, since μ and the change of δ are always in the same direction, and since μ depends on the standard T_o only, while δ depends on latitude, the apparatus can be constructed so that for some one definite latitude (λ_1) the ballistic deflection shall amount exactly to the difference between the angle δ of the old course and the angle δ of the new course, so that the Compass comes immediately to its new resting position, quite " dead-beat."

For this purpose it is merely necessary to arrange for a period of oscillation of appropriate length.

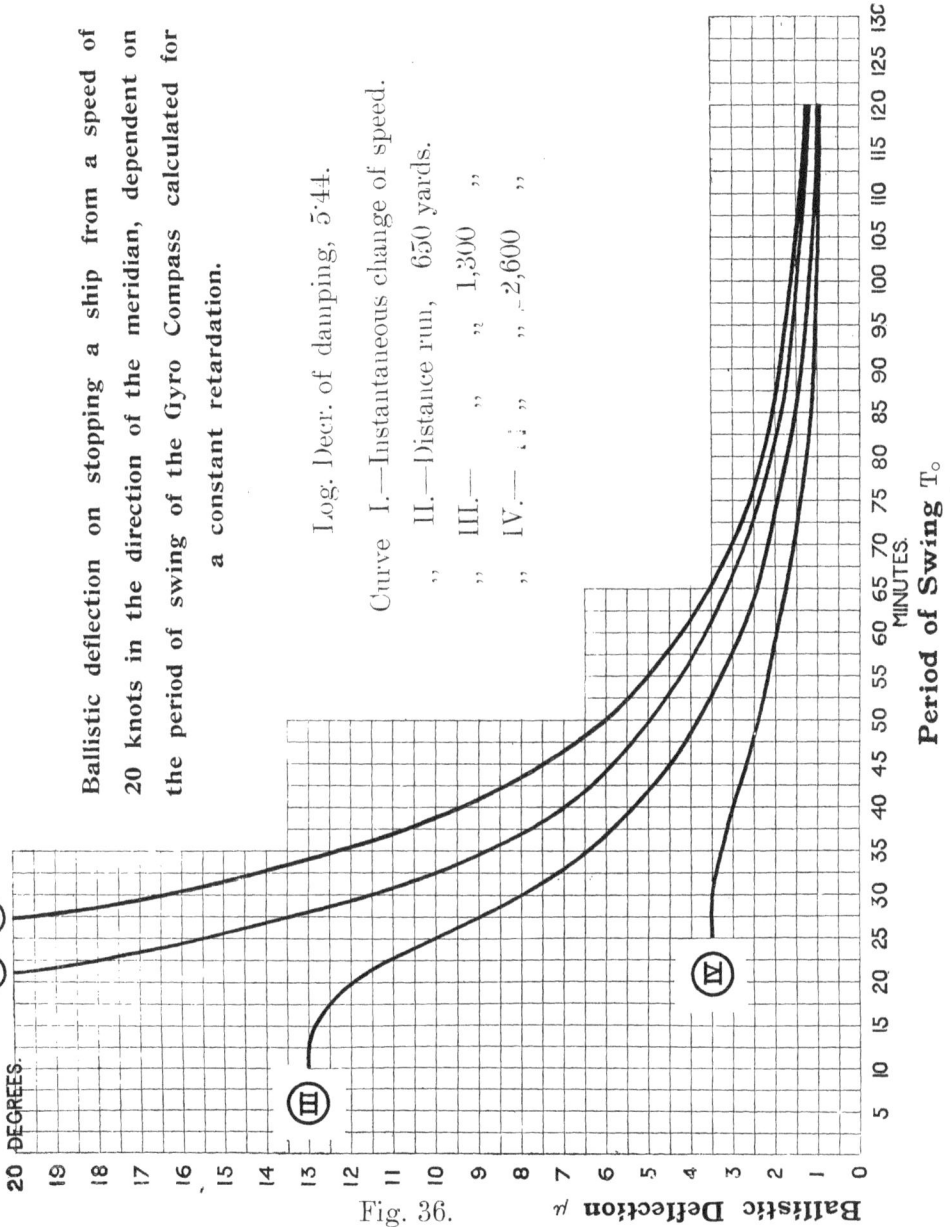

Ballistic deflection on stopping a ship from a speed of 20 knots in the direction of the meridian, dependent on the period of swing of the Gyro Compass calculated for a constant retardation.

Log. Decr. of damping, 5·44.

Curve I.—Instantaneous change of speed.
,, II.—Distance run, 650 yards.
,, III.— ,, ,, 1,300 ,,
,, IV.— ,, ,, 2,600 ,,

Fig. 36.

Period of Swing T₀

Ballistic Deflection μ

F

To find what this must be, we must equate the value of μ, viz. :—

$$\frac{4\ \pi^2}{g\ \omega\ \cos\lambda_o}\ .\ \frac{1}{T_o^2}\ (v_2 - v_1)$$

to the change in δ, which is $\dfrac{v_2 - v_1}{E\ \omega\ \cos\lambda_1}$ (see equation (17), page 69).

Hence $T_o^2 = \dfrac{4\ \pi^2\ E\ \cos\lambda_1}{g\ \cos\lambda_o}$

i.e., the standard value of T_o must be $2\ \pi\ \sqrt{\dfrac{E\ \cos\lambda_1}{g\ \cos\lambda_o}}$

It is most interesting to notice that the actual period of undamped oscillations at the latitude λ_1 for which the ballistic deflections are dead-beat (here denoted by T to distinguish it from the standard value T_o corresponding to latitude λ_o) is :

$$(22) \qquad\qquad T = 2\ \pi\ \sqrt{\frac{E}{g}}$$

since from equation (5), page 57,

$$T^2 \cos\lambda_1 = T_o^2 \cos\lambda_o \left[= \frac{4\ \pi^2\ H}{M\ g\ a\ \omega} \right]$$

that is to say :—

At any latitude in which the ballistic deflections are "dead-beat," the actual period of undamped precessional oscillations must be equal to the period of oscillation of a simple pendulum, whose length equals the earth's radius : this is about 85 minutes.

In constructing the instrument we can arrange the standard value of T_o, so that the actual period may have the above value at any desired latitude λ_1.

The following table gives the various standard values of T_o (either for different instruments or for different adjustments of the same instrument), and the corresponding values of μ ; and in the last column the range of latitude is indicated within which the movements of the Gyro will be practically "dead-beat"; that is to say, in which the actual oscillations will approximately synchronize with a pendulum of length E, which is the condition which makes μ equal to the **change** in δ.

In other latitudes μ disappears by damping according to the curve Figure 33, page 66.

Maximum Ballistic Deflection μ.

Dependent on the undamped period of oscillation at λ_o
Alterations of speed in direction of the meridian.

SPEED IN KNOTS.

T_o	2	4	6	8	10	12	14	16	18	20	22	24	26	28	
5	62.0	90	—	—	—	—	—	—	—	—	—	—	—	—	
10	15.5	31.0	46.5	62.0	77.6	90	—	—	—	—	—	—	—	—	
20	3.9	7.8	11.6	15.5	19.4	23.3	27.2	31.0	34.9	38.8	42.7	46.5	50.4	54.2	
30	1.7	3.5	5.2	6.9	8.6	10.4	12.1	13.8	15.5	17.3	19.0	20.7	22.4	24.2	
40	1.0	1.9	2.9	3.9	4.9	5.8	6.8	7.8	8.7	9.7	10.7	11.6	12.6	13.6	
45	0.8	1.5	2.3	3.1	3.8	4.6	5.4	6.1	6.9	7.7	8.4	9.2	10.0	10.7	
50	0.6	1.2	1.9	2.5	3.1	3.7	4.4	5.0	5.6	6.2	6.8	7.5	8.1	8.7	
55	0.5	1.0	1.5	2.0	2.6	3.1	3.6	4.1	4.6	5.1	5.6	6.1	6.7	7.2	$\mu = \delta$
60	0.4	0.9	1.3	1.7	2.2	2.6	3.0	3.5	3.9	4.3	4.7	5.2	5.6	6.0	in latitudes of
65	0.4	0.7	1.1	1.5	1.8	2.2	2.6	2.5	3.3	3.7	4.0	4.4	4.8	5.1	70°
70	0.3	0.6	1.0	1.3	1.6	1.9	2.2	2.5	2.9	3.2	3.5	3.8	4.1	4.4	
75	0.3	0.6	0.8	1.1	1.4	1.7	1.9	2.2	2.5	2.8	3.0	3.3	3.6	3.9	60°
80	0.2	0.5	0.7	1.0	1.2	1.5	1.7	1.9	2.2	2.4	2.7	2.9	3.2	3.4	55°—60°
90	0.2	0.4	0.6	0.8	1.0	1.2	1.3	1.5	1.7	1.9	2.1	2.3	2.5	2.7	45°—55°
100	0.2	0.3	0.5	0.6	0.8	0.9	1.1	1.2	1.4	1.6	1.7	1.9	2.0	2.2	30°—45°
110	0.1	0.3	0.4	0.5	0.6	0.8	0.9	1.0	1.2	1.3	1.4	1.5	1.7	1.8	0°—30°
120	0.1	0.2	0.3	0.4	0.5	0.6	0.8	0.9	1.0	1.1	1.2	1.3	1.4	1.5	

(Left axis: Undamped Period of Oscillation in Minutes.)

In practice, sudden changes in the ship's speed do not occur, and consequently the observed values of the ballistic deflection will be smaller than the maximum values which can be calculated from equation (21). In order to get a nearer approximation to the real conditions we must make certain assumptions as to the nature of the acceleration.

The curves of Figure 36, page 73, are calculated for the simplest case, where γ is constant. The method of obtaining the solution for this simple case is best realized by considering the effect of adding the turning couple $M a \gamma$ in equation (1), page 55, and referring to the diagram in Figure 33, page 66, where the motion of the Gyro in both planes is depicted.

While the Gyro is steady, $a = a_o$ and $\beta = \beta_o$.

If now the acceleration γ acts, equation (1) becomes :—

$$(23) \quad H \left(\frac{d\,a}{d\,t} + \omega \sin \lambda \right) = M\,g\,a\,\beta + M\,a\,\gamma$$

$$= M\,g\,a \left(\beta + \frac{\gamma}{g} \right)$$

Hence the value of β which now corresponds to $\frac{d\,a}{d\,t} = 0$

is $\beta_o - \frac{\gamma}{g}$ instead of β_o; that is to say, the axis is above its position of equilibrium, and swings towards it under the increased turning couple, in a definite elliptical spiral, Figure 33, page 66, which spiral is followed so long as the acceleration γ is maintained.

When the ship again proceeds on its course with a constant speed, the acceleration γ disappears, the expression $\frac{\gamma}{g}$ is zero, and the axis moves to its resting position, which is shifted to the value of δ for the new course, the Gyro swinging into this position with oscillations according to the curve of Figure 33, page 66.

Table of Symbols employed in the above Investigation.

a   Metacentric height.

β Angle of inclination of Gyro axle to the horizontal.

a Azimuthal angle between north end of Gyro and north end of meridian (positive to the west).

M   Mass of movable parts of the instrument.

ω Angular velocity of earth's rotation.

λ Angle of latitude.

H   Angular momentum of Gyro wheel about its axis.

I   Moment of inertia of imaginary pendulum isochronous with the system $\left(= \dfrac{H^2}{M\, g\, a} \right)$

R   The righting co-efficient of the Gyro. $(= H\, \omega \cos \lambda)$

T_o Time of one complete oscillation of undamped Gyro $\left(= 2\,\pi\sqrt{\dfrac{I}{R}} = 2\,\pi\sqrt{\dfrac{H}{M\, g\, a\, \omega \cos \lambda}} \right)$

D   Co-efficient of air blast, D β.

k   Co-efficient of retardation $= \left(\dfrac{H\, D}{M\, g\, a} \right)$

a_o Value of a in resting position, $i.e.$, latitude correction.

β_o Value of β when $\dfrac{d\, a}{d\, t} = 0$.

A   Constant of amplitude.

T_1 Time of one complete oscillation of damped Gyro.

p $= \dfrac{k}{2\, I}$

ϵ Numerical value of logarithmic decrement.

v Component of ship's speed in N–S direction.

E   Radius of the earth.

δ Change in azimuthal angle due to speed of ship.

γ Acceleration of ship in N–S direction.

μ Ballistic deflection.

λ_o Latitude for which Gyro Compass is standardized.

λ_1 Latitude for which ballistic deflection of Compass is "dead-beat."

Equations employed in the above Investigations.

No Damping.

(1) $\qquad H \left(\dfrac{d\,a}{d\,t} + \omega \sin \lambda\right) = M\,g\,a\,\beta$

(2) $\qquad H \left(\dfrac{d\,\beta}{d\,t} + \omega \cos \lambda \sin a\right) = 0$

(3) $\qquad H \dfrac{d^2\,a}{d\,t^2} = M\,g\,a\,\dfrac{d\,\beta}{d\,t}$ by differentiating (1)

(4) $\qquad \dfrac{H^2}{M\,g\,a} \cdot \dfrac{d^2\,a}{d\,t^2} + H\,\omega \cos \lambda \sin a = 0$

$\qquad$ *i.e.* $\quad I \dfrac{d^2\,a}{d\,t^2} + R \sin a = 0$

(5) $\qquad T_o = 2\,\pi \sqrt{\dfrac{I}{R}} = 2\,\pi \sqrt{\dfrac{H}{M\,g\,a\,\omega \cos \lambda}}$

With Damping.

(6) $\qquad \begin{cases} \text{(i.)} \ H \left(\dfrac{d\,a}{d\,t} + \omega \sin \lambda\right) = M\,g\,a\,\beta \\[2ex] \text{(ii.)} \ H \left(\dfrac{d\,\beta}{d\,t} + \omega \cos \lambda \sin a\right) = -\,D\,\beta \end{cases}$

(6a) $\qquad \begin{cases} \dfrac{H^2}{M\,ga} \cdot \dfrac{d^2\,a}{d\,t^2} + H\,\omega \cos \lambda \sin a + \dfrac{H\,D}{M\,ga}\left(\dfrac{d\,a}{d\,t} + \omega \sin \lambda\right) = 0 \\[2ex] \text{\textit{i.e.}} \ I \dfrac{d^2\,a}{d\,t^2} + R \sin a + k \left(\dfrac{d\,a}{d\,t} + \omega \sin \lambda\right) = 0 \end{cases}$

whence $\dfrac{k}{H} = - \dfrac{\dfrac{d\,\beta}{d\,t} + \omega\,\cos\,\lambda\,\sin\,a}{\dfrac{d\,a}{d\,t} + \omega\,\sin\,\lambda} = -\tan\,\phi$

where ϕ is the angle made with the vertical by the axis of total angular velocity of the Gyro axle.

(7) $\qquad H\,\omega\,\cos\,\lambda\,\sin\,a_\circ = -\dfrac{H\,D}{M\,g\,a}\,\omega\,\sin\,\lambda$

$\qquad\qquad i.e.\ \ a_\circ = -\dfrac{D}{M\,g\,a}\,\tan\,\lambda$

(8) $\qquad \beta_\circ\left(\dfrac{i.e.\ \beta \text{ when}}{\dfrac{d\,a}{d\,t} = 0}\right) = \dfrac{H\,\omega\,\sin\,\lambda}{M\,g\,a}$

(9) $\qquad I\,\dfrac{d^2\,a_1}{d\,t^2} + k\,\dfrac{d\,a_1}{d\,t} + R\,a_1 = 0$

(10) $\quad a_1 = a - a_\circ = A\,e^{-\dfrac{k\,t}{2\,I}}\sin\,\sqrt{\dfrac{R}{I} - \dfrac{k^2}{4\,I^2}}\cdot t$

(11) $\qquad\qquad T_1 = \dfrac{4\,\pi\,I}{\sqrt{4\,I\,R - k^2}}$

(11a) $\qquad a = A\,e^{-\dfrac{k\,t}{2\,I}}\sin\,\dfrac{2\,\pi\,t}{T_1} + a_\circ$

$-\log.\,(\text{dec.}) = \epsilon = \dfrac{k\,T_1}{4\,I} = \dfrac{D\,T_1}{4\,H} = \pi\,\cot\,\eta = \pi\,\tan\,\xi$

No Damping.

(12) $\qquad \begin{cases} a = A\,\sin\,\dfrac{2\,\pi\,t}{T_\circ} \\[2ex] \beta = A\,\dfrac{2\,\pi}{T_\circ}\cdot\dfrac{H}{M\,g\,a}\,\cos\,\dfrac{2\,\pi\,t}{T_\circ} + \beta_\circ \end{cases}$

With Damping.

$$(13)\begin{cases} a = A\, e^{-\frac{k\,t}{2\,I}}\sin\frac{2\,\pi\,t}{T_1} + a_\circ \\[2mm] \beta = \frac{H}{M\,g\,a}\cdot\frac{d\,a}{d\,t} + \beta_\circ \\[2mm] = \frac{H}{M\,g\,a}\cdot A\,e^{-\frac{k\,t}{2\,I}}\left(\frac{2\,\pi}{T_1}\cos\frac{2\,\pi\,t}{T_1} - \frac{k}{2\,I}\sin\frac{2\,\pi\,t}{T_1}\right) + \beta_\circ \end{cases}$$

$$(14)\begin{cases} \left(\frac{2\,\pi}{T_1}\right)^2 + \left(\frac{k}{2\,I}\right)^2 = \left(\frac{2\,\pi}{T_\circ}\right)^2 \\[2mm] \left(\frac{T_1}{T_\circ}\right)^2 = 1 + \left(\frac{\epsilon}{\pi}\right)^2 \end{cases}$$

$$(15)\begin{cases} \frac{2\,\pi}{T_\circ}\cos\xi = \frac{2\,\pi}{T_1} \\[2mm] \frac{2\,\pi}{T_\circ}\sin\xi = \frac{k}{2\,I} \end{cases}$$

$$(16)\begin{cases} a = A\, e^{-\frac{k\,t}{2\,I}}\sin\frac{2\,\pi\,t}{T_1} + a_\circ \\[2mm] \beta = \frac{A\,H}{M\,g\,a}\cdot\frac{2\,\pi}{T_\circ}\,e^{-\frac{k\,t}{2\,I}}\cos\left(\frac{2\,\pi\,t}{T_1} + \xi\right) + \beta_\circ \end{cases}$$

Deflections.   Constant Speed.

$$(17)\qquad \delta = \frac{v}{E}\cdot\frac{1}{\omega\,\cos\lambda}$$

81

Under Acceleration.

$$(18) \qquad H \frac{d\,a}{d\,t} = M\,a\,\gamma$$

$$(19) \qquad \mu = \frac{M\,a}{H} \int_0^T \gamma\,d\,t$$

$$(20) \qquad \mu = \frac{M\,a}{H} (v_2 - v_1)$$

$$(21) \quad \begin{cases} \mu = \frac{4\,\pi^2}{g\,\omega\,\cos\,\lambda_0} \cdot \frac{1}{T_c^2} (v_2 - v_1) \\ \text{where } T_0 \text{ is the period of an undamped oscillation} \\ \text{at latitude } \lambda_0. \end{cases}$$

$$(22) \quad \begin{cases} \text{When the ballistic deflections are "dead-beat,"} \\ \text{the period of an undamped oscillation is} \\ T = 2\,\pi\,\sqrt{\frac{E}{g}} \end{cases}$$

$$(23) \quad H \left(\frac{d\,a}{d\,t} + \omega\,\sin\,\lambda \right) = M\,g\,a \left(\beta + \frac{\gamma}{g} \right)$$

Practical Use on Board Ship.

In the use of the Gyro Compass **two** corrections have to be considered.

    1. The **latitude correction** dependent upon the latitude only.

    2. The **angle** δ which depends upon the latitude, speed and course of the ship.

(The complete analysis of these two corrections is given in the chapter on Theory.)

The reason for the **first** correction arises from the fact that the directive power is at a **maximum** at the equator and a **minimum** at the poles, varying proportionally between these two positions. The effect of this is to change slightly the position of rest of the Compass Card in different latitudes, as explained more fully at pages 35–38.

In order to eliminate this correction from the reading of the **single instrument**, the lubber line is moveable. The whole of the cover being held in position by two set screws, if the ship goes into some latitude other than 50° N (this being the latitude for which the Compass is adjusted) these two screws may be loosened and the adjustment on the lubber line shifted by the amount given in the table below :—

Latitude 60° north,	·6 (	36′) easterly.
,, 50° ,,		
,, 40° ,,	·5 (	30′) westerly.
,, 20° ,,	1°·1 (1°	6′) ,,
,, 0° ,,	1°·6 (1°	36′) ,,
,, 20° south	2°·1 (2°	6′) ,,
,, 40° ,,	2°·7 (2°	42′) ,,
,, 60° ,,	3°·8 (3°	48°) ,,

In making this correction the errors are treated in the same way as the deviations of the Magnetic Compass.

Example :—If in latitude 60° north the ship's head is N 32° E (32°) by the Gyro Compass, then the real course is :—

North 32° east.   (32°)

add _____ ·6 east.   (Latitude correction.)

corrected result north 32°·6 east. (32°·6)

or north 32°   36′ east.

From the table above it may be seen that the first of the two corrections has only to be used when a change of latitude of 10° has occurred.

It should be noted that alteration of the position of the lubber line makes the actual Gyro Compass correct for steering by, but if the **single Compass** itself were used for taking bearings, the correction for latitude would have to be taken into account, as the N and S line on the card would no longer correspond exactly with the meridian.

When the **transmission system** is installed, the position of the lubber line can be shifted in the Master Compass and **each individual receiver** can be adjusted so that the correct point on the card corresponds with the lubber line, as in the case of the Master Compass card and its lubber line in its new position; and by this means the **receivers** can be made correct for steering, and **also** for taking **bearings.**

A very simple means is provided in the receivers for setting their indications, a cover at one side of the case can be removed, and a small stem can then be turned, the amount of turning being seen by the movement of the central Compass Card.

Fig. 37.

Gyro Compass with Top Cover Removed.

a Level.
b Central Stem.
c Clamp screw for lubber point.

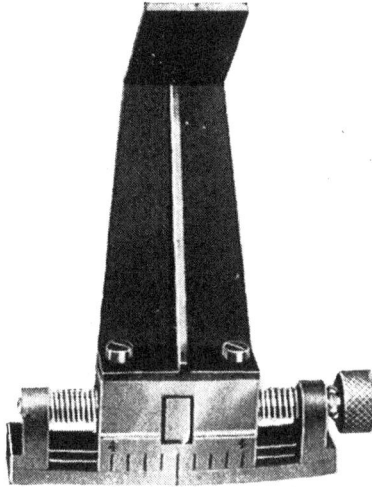

Fig. 38.

Figure 38 illustrates the adjustment actually provided in the Master Compass for adjusting the lubber line; this can be got at readily when the top cover is opened.

The **second** correction for the angle δ is dependent on the **latitude, the ship's speed**, and her **course**; the values are given in the Tables, pages 89-92. The amount of this correction is far smaller than is the case with the Magnetic Compass.

In manœuvring, and when the ship's speed is low, this correction may be neglected altogether.

Variations in the pressure of the ship's direct current lighting circuit are of no account unless they are so great and last for so long that the speed of rotation of the Gyro is **very** considerably changed.

Testing.

It is recommended, when endeavouring to obtain the very highest degree of accuracy for observations, that about eight readings should be taken at times about five minutes apart; then, if these are plotted graphically, the very smallest amount of swinging movement can be detected, if the compass has not settled down to its final reading. The Gyro should have been running for at least three hours before such tests are made.

When several readings are taken in close succession, and one such reading differs from the others, it will be found in most cases that this difference is due to an error in reading, or to an error in the apparatus with which the Gyro Compass is being compared.

The Gyro Compass is independent of magnetic influences and therefore neither **variation** nor **deviation** have to be taken into account. If the direction is correct for one course it is correct for all courses.

As a **practical example** let us consider the case of a newly installed Gyro Compass being checked while the ship steams **18 miles, N.W.** $(315°)$ in **one hour** on a **straight course**, checked by means of bearings on objects on land; in latitude $50°$ north. The course can be checked as being correct at the instants when the readings are taken.

Suppose the Gyro Compass readings are :—

$(317°·8)$
$(317°·7)$
$(317°·1)$
$(322°·4)$
$(317°·8)$
$(317°·0)$
$(318°\ \ )$
$(317°·5)$

We can say with confidence that the fourth reading has an error of observation of some $5°$ too high ; and it can also be seen by the uniformity of the readings that no swinging of the compass has taken place, or, in other words, that it has settled down. The difference in the decimals of a degree are due to reading errors in taking the bearings.

The figures should be added, and the 5° error in the fourth reading deducted, thus

$$2545°{\cdot}3$$
$$5°$$
$$\overline{2540°{\cdot}3}$$

Dividing this total by 8 (the number of the readings) we obtain

$$8 \,) \, \underline{2540°{\cdot}3}$$
$$317°{\cdot}5$$

317°·5 is therefore the mean reading. A correction must be applied to this, due to angle δ, the value for which is given in the table, page 89. First look at the table for latitude 50°, then on the horizontal line corresponding to a course north-west (315°), the value of 1·3 is seen in the column for a speed of 18 knots.

$$317{\cdot}5$$
$$\text{Deduct} \quad 1{\cdot}3$$
$$\text{Course by Gyro} = \overline{316{\cdot}2}$$

The correct bearing of the course should be 315°, as it will be remembered it was due north-west. The Gyro Compass would therefore appear to be 1°·2 out, and this error can be taken out by moving the lubber line on the Gyro Compass this amount to the west.

By doing so the Gyro Compass is adjusted correctly for **all courses.**

On a passage from Lisbon to New York no error due to change of latitude need be taken into account, and the angle δ can be neglected on account of the course being almost exactly west. In a passage from Plymouth to Gibraltar the angle δ would have to be considered as well as the error due to change of latitude set out on page 82.

Table for the Angle δ.

The correction for the angle δ is applied like a deviation correction for a Magnetic Compass.

The two columns, δ westerly and δ easterly, indicate the direction of the corrections for various courses.

The corrections are given in degrees and decimals of a degree.

It will be noticed that the worst condition which can be met with in ordinary navigable latitudes at ordinary speeds is at latitude 60° north, with a speed of 24 knots, on a course either due north (when 3°·1 westerly correction is required) or a course due south at the same speed, when the correction is 3°·1 easterly.

The figures in brackets are the actual readings as marked on the Compass Cards, which are marked 0 to 360°.

Table for Angle δ, in Degrees and Decimals.

LATITUDE 0°

δ Westerly Course		δ Easterly Course		6	8	10	12	14	16	18	20	22	24	26	28	30
N (0°)	N (0°)	S (180°)	S (180°)	0.4	0.5	0.6	0.8	0.9	1.0	1.1	1.3	1.4	1.5	1.6	1.7	1.9
N. 10°E (10°)	N. 10°W (350°)	S. 10°E (170°)	S. 10°W (190°)	0.4	0.5	0.6	0.7	0.8	1.0	1.1	1.2	1.4	1.5	1.6	1.7	1.8
N. 20°E (20°)	N. 20°W (340°)	S. 20°E (160°)	S. 20°W (200°)	0.4	0.5	0.6	0.7	0.8	0.9	1.1	1.2	1.3	1.4	1.5	1.6	1.7
N. 30°E (30°)	N. 30°W (330°)	S. 30°E (150°)	S. 30°W (210°)	0.3	0.4	0.5	0.6	0.7	0.8	1.0	1.0	1.1	1.3	1.4	1.5	1.6
N. 40°E (40°)	N. 40°W (320°)	S. 40°E (140°)	S. 40°W (220°)	0.3	0.4	0.5	0.5	0.6	0.7	0.9	1.0	1.1	1.2	1.3	1.4	1.4
N. 50°E (50°)	N. 50°W (310°)	S. 50°E (130°)	S. 50°W (230°)	0.3	0.3	0.4	0.5	0.6	0.7	0.7	0.8	0.9	1.0	1.1	1.1	1.2
N. 60°E (60°)	N. 60°W (300°)	S. 60°E (120°)	S. 60°W (240°)	0.2	0.2	0.3	0.4	0.5	0.5	0.5	0.7	0.7	0.8	0.8	0.9	1.0
N. 70°E (70°)	N. 70°W (290°)	S. 70°E (110°)	S. 70°W (250°)	0.1	0.1	0.2	0.3	0.3	0.4	0.4	0.4	0.5	0.5	0.6	0.6	0.6
N. 80°E (80°)	N. 80°W (280°)	S. 80°E (100°)	S. 80°W (260°)	0.1	—	0.1	0.1	0.1	0.2	0.2	0.2	0.3	0.3	0.3	0.3	0.3
E (90°)	W (270°)	E (90°)	W (270°)	—	—	—	—	—	—	—	—	—	—	—	—	—

Speed in Knots.

LATITUDE 10° (NORTH OR SOUTH).

δ Westerly Course		δ Easterly Course		6	8	10	12	14	16	18	20	22	24	26	28	30
N (0°)	N (0°)	S (180°)	S (180°)	0.4	0.5	0.6	0.8	0.9	1.0	1.2	1.3	1.4	1.5	1.6	1.7	1.9
N. 10°E (10°)	N. 10°W (350°)	S. 10°E (170°)	S. 10°W (190°)	0.4	0.5	0.6	0.8	0.9	1.0	1.1	1.3	1.4	1.5	1.6	1.7	1.8
N. 20°E (20°)	N. 20°W (340°)	S. 20°E (160°)	S. 20°W (200°)	0.4	0.5	0.6	0.7	0.8	1.0	1.1	1.2	1.3	1.4	1.5	1.7	1.8
N. 30°E (30°)	N. 30°W (330°)	S. 30°E (150°)	S. 30°W (210°)	0.3	0.4	0.5	0.6	0.8	0.9	1.0	1.1	1.2	1.3	1.4	1.6	1.7
N. 40°E (40°)	N. 40°W (320°)	S. 40°E (140°)	S. 40°W (220°)	0.3	0.4	0.5	0.5	0.7	0.8	0.9	1.0	1.1	1.2	1.3	1.4	1.4
N. 50°E (50°)	N. 50°W (310°)	S. 50°E (130°)	S. 50°W (230°)	0.3	0.3	0.3	0.4	0.6	0.7	0.7	0.8	0.9	1.0	1.1	1.1	1.2
N. 60°E (60°)	N. 60°W (300°)	S. 60°E (120°)	S. 60°W (240°)	0.2	0.2	0.3	0.3	0.4	0.5	0.5	0.6	0.7	0.8	0.8	0.9	1.0
N. 70°E (70°)	N. 70°W (290°)	S. 70°E (110°)	S. 70°W (250°)	0.1	0.1	0.2	0.2	0.3	0.4	0.4	0.5	0.5	0.6	0.6	0.6	0.6
N. 80°E (80°)	N. 80°W (280°)	S. 80°E (100°)	S. 80°W (260°)	—	—	0.1	0.1	0.2	0.2	0.2	0.3	0.3	0.3	0.3	0.3	0.4
E (90°)	W (270°)	E (90°)	W (270°)	—	—	—	—	—	—	—	—	—	—	—	—	—

Speed in Knots.

Table for Angle δ, in Degrees and Decimals.

LATITUDE 20° (NORTH OR SOUTH).

δ Westerly Course		δ Easterly Course		Speed in Knots												
				6	8	10	12	14	16	18	20	22	24	26	28	30
N (0°)	N (0°)	S (180°)	S (180°)	0.4	0.5	0.7	0.8	0.9	1.1	1.2	1.4	1.5	1.6	1.7	1.8	2.0
N.10°E (10°)	N.10°W (350°)	S.10°E (170°)	S.10°W (190°)	0.4	0.5	0.7	0.8	0.9	1.1	1.2	1.3	1.5	1.6	1.7	1.8	2.0
N.20°E (20°)	N.20°W (340°)	S.20°E (160°)	S.20°W (200°)	0.4	0.5	0.6	0.8	0.9	1.0	1.1	1.3	1.4	1.5	1.6	1.7	1.9
N.30°E (30°)	N.30°W (330°)	S.30°E (150°)	S.30°W (210°)	0.3	0.5	0.6	0.7	0.8	0.9	1.0	1.1	1.3	1.4	1.5	1.6	1.7
N.40°E (40°)	N.40°W (320°)	S.40°E (140°)	S.40°W (220°)	0.3	0.4	0.5	0.6	0.7	0.8	0.9	1.0	1.1	1.3	1.4	1.5	1.6
N.50°E (50°)	N.50°W (310°)	S.50°E (130°)	S.50°W (230°)	0.2	0.3	0.5	0.5	0.6	0.7	0.8	0.9	0.9	1.1	1.2	1.2	1.3
N.60°E (60°)	N.60°W (300°)	S.60°E (120°)	S.60°W (240°)	0.2	0.3	0.4	0.4	0.5	0.5	0.6	0.7	0.7	0.8	0.9	0.9	1.0
N.70°E (70°)	N.70°W (290°)	S.70°E (110°)	S.70°W (250°)	0.2	0.2	0.3	0.3	0.4	0.4	0.4	0.5	0.5	0.5	0.6	0.6	0.7
N.80°E (80°)	N.80°W (280°)	S.80°E (100°)	S.80°W (260°)	0.1	0.1	0.1	0.2	0.2	0.2	0.2	0.2	0.3	0.3	0.3	0.3	0.3
E (90°)	W (270°)	E (90°)	W (270°)	—	—	—	—	—	—	—	—	—	—	—	—	—

LATITUDE 30° (NORTH OR SOUTH).

δ Westerly Course		δ Easterly Course		Speed in Knots												
				6	8	10	12	14	16	18	20	22	24	26	28	30
N (0°)	N (0°)	S (180°)	S (180°)	0.4	0.6	0.7	0.9	1.0	1.2	1.3	1.5	1.6	1.7	1.9	2.0	2.1
N.10°E (10°)	N.10°W (350°)	S.10°E (170°)	S.10°W (190°)	0.4	0.6	0.7	0.9	1.0	1.1	1.3	1.4	1.6	1.7	1.8	1.9	2.1
N.20°E (20°)	N.20°W (340°)	S.20°E (160°)	S.20°W (200°)	0.4	0.5	0.7	0.8	0.9	1.1	1.2	1.4	1.5	1.6	1.6	1.7	1.9
N.30°E (30°)	N.30°W (330°)	S.30°E (150°)	S.30°W (210°)	0.4	0.5	0.6	0.7	0.9	1.0	1.1	1.2	1.4	1.5	1.6	1.7	1.8
N.40°E (40°)	N.40°W (320°)	S.40°E (140°)	S.40°W (220°)	0.3	0.4	0.5	0.6	0.8	0.9	1.0	1.1	1.2	1.3	1.4	1.5	1.7
N.50°E (50°)	N.50°W (310°)	S.50°E (130°)	S.50°W (230°)	0.3	0.3	0.4	0.6	0.7	0.8	0.8	0.9	1.0	1.1	1.1	1.2	1.3
N.60°E (60°)	N.60°W (300°)	S.60°E (120°)	S.60°W (240°)	0.2	0.3	0.4	0.4	0.5	0.6	0.6	0.7	0.8	0.9	0.9	0.9	1.1
N.70°E (70°)	N.70°W (290°)	S.70°E (110°)	S.70°W (250°)	0.2	0.2	0.3	0.3	0.4	0.4	0.5	0.5	0.6	0.6	0.6	0.6	0.7
N.80°E (80°)	N.80°W (280°)	S.80°E (100°)	S.80°W (260°)	0.1	0.1	0.1	0.2	0.2	0.2	0.3	0.3	0.3	0.3	0.3	0.3	0.3
E (90°)	W (270°)	E (90°)	W (270°)	—	—	—	—	—	—	—	—	—	—	—	—	—

Table for Angle δ, in Degrees and Decimals.

LATITUDE 40° (NORTH OR SOUTH).

δ Westerly Course	δ Easterly Course	δ Easterly Course	6	8	10	12	14	16	18	20	22	24	26	28	30
N (0°)	S (180°)	S (180°)	0.5	0.7	0.8	1.0	1.2	1.3	1.5	1.7	1.8	2.0	2.1	2.3	2.5
N. 10°W (350°)	S. 10°E (170°)	S. 10°W (190°)	0.5	0.7	0.8	1.0	1.1	1.3	1.5	1.6	1.8	2.0	2.1	2.3	2.4
N. 20°W (340°)	S. 20°E (160°)	S. 20°W (200°)	0.5	0.6	0.7	0.9	1.1	1.2	1.4	1.5	1.7	1.8	1.9	2.1	2.2
N. 30°W (330°)	S. 30°E (150°)	S. 30°W (210°)	0.4	0.6	0.7	0.8	1.0	1.1	1.3	1.4	1.6	1.7	1.8	1.9	2.0
N. 40°W (320°)	S. 40°E (140°)	S. 40°W (220°)	0.4	0.5	0.6	0.7	0.9	1.0	1.1	1.3	1.4	1.5	1.6	1.7	1.8
N. 50°W (310°)	S. 50°E (130°)	S. 50°W (230°)	0.4	0.5	0.6	0.6	0.8	0.8	1.0	1.1	1.2	1.3	1.3	1.4	1.5
N. 60°W (300°)	S. 60°E (120°)	S. 60°W (240°)	0.3	0.4	0.4	0.5	0.6	0.7	0.7	0.8	0.9	1.0	1.1	1.1	1.2
N. 70°W (290°)	S. 70°E (110°)	S. 70°W (250°)	0.2	0.3	0.3	0.4	0.4	0.5	0.5	0.5	0.7	0.7	0.7	0.8	0.9
N. 80°W (280°)	S. 80°E (100°)	S. 80°W (260°)	0.1	0.1	0.2	0.2	0.2	0.3	0.3	0.3	0.4	0.4	0.4	0.4	0.5
W (270°)	E (90°)	W (270°)	—	—	—	—	—	—	—	—	—	—	—	—	—

(Speed in Knots.)

LATITUDE 50° (NORTH OR SOUTH).

δ Westerly Course	δ Easterly Course	δ Easterly Course	6	8	10	12	14	16	18	20	22	24	26	28	30
N (0°)	S (180°)	S (180°)	0.6	0.8	1.0	1.2	1.4	1.6	1.8	2.0	2.2	2.4	2.6	2.8	3.0
N. 10°W (350°)	S. 10°E (170°)	S. 10°W (190°)	0.6	0.8	1.0	1.2	1.4	1.5	1.7	1.9	2.1	2.3	2.5	2.7	2.9
N. 20°W (340°)	S. 20°E (160°)	S. 20°W (200°)	0.5	0.7	0.9	1.1	1.3	1.5	1.6	1.8	2.0	2.2	2.4	2.5	2.7
N. 30°W (330°)	S. 30°E (150°)	S. 30°W (210°)	0.5	0.7	0.8	1.0	1.2	1.4	1.5	1.7	1.9	2.1	2.2	2.3	2.5
N. 40°W (320°)	S. 40°E (140°)	S. 40°W (220°)	0.4	0.6	0.7	0.9	1.1	1.2	1.4	1.5	1.6	1.8	1.9	2.1	2.3
N. 50°W (310°)	S. 50°E (130°)	S. 50°W (230°)	0.4	0.5	0.7	0.8	0.9	1.0	1.2	1.3	1.4	1.5	1.6	1.7	1.9
N. 60°W (300°)	S. 60°E (120°)	S. 60°W (240°)	0.3	0.4	0.5	0.6	0.7	0.8	1.0	1.0	1.1	1.2	1.2	1.3	1.4
N. 70°W (290°)	S. 70°E (110°)	S. 70°W (250°)	0.2	0.3	0.4	0.5	0.5	0.6	0.7	0.7	0.8	0.9	0.9	0.9	1.0
N. 80°W (280°)	S. 80°E (100°)	S. 80°W (260°)	0.1	0.2	0.2	0.2	0.3	0.3	0.3	0.4	0.4	0.5	0.5	0.5	0.5
W (270°)	E (90°)	W (270°)	—	—	—	—	—	—	—	—	—	—	—	—	—

(Speed in Knots.)

Table for Angle δ, in Degrees and Decimals.

LATITUDE 60° (NORTH OR SOUTH).

δ Westerly. Course.		δ Easterly. Course.		Speed in Knots.												
				6	8	10	12	14	16	18	20	22	24	26	28	30
N. (0°)	N. (0°)	S (180°)	S (180°)	0.8	1.0	1.3	1.5	1.8	2.0	2.2	2.5	2.8	3.1	3.3	3.5	3.8
N. 10°E (10°)	N. 10°W (350°)	S. 10°E (170°)	S. 10°W (190°)	0.8	1.0	1.3	1.5	1.7	2.0	2.0	2.5	2.7	3.0	3.3	3.5	3.7
N. 20°E (20°)	N. 20°W (340°)	S. 20°E (160°)	S. 20°W (200°)	0.7	0.9	1.2	1.4	1.6	1.9	2.1	2.3	2.6	2.8	3.0	3.2	3.5
N. 30°E (30°)	N. 30°W (330°)	S. 30°E (150°)	S. 30°W (210°)	0.6	0.8	1.1	1.3	1.5	1.8	2.0	2.2	2.4	2.6	2.8	3.0	3.2
N. 40°E (40°)	N. 40°W (320°)	S. 40°E (140°)	S. 40°W (220°)	0.5	0.7	1.0	1.2	1.4	1.5	1.7	1.9	2.1	2.3	2.5	2.7	2.9
N. 50°E (50°)	N. 50°W (310°)	S. 50°E (130°)	S. 50°W (230°)	0.5	0.7	0.8	1.0	1.2	1.3	1.5	1.6	1.8	2.0	2.1	2.3	2.4
N. 60°E (60°)	N. 60°W (300°)	S. 60°E (120°)	S. 60°W (240°)	0.4	0.5	0.6	0.8	0.9	1.0	1.2	1.2	1.5	1.6	1.6	1.8	1.9
N. 70°E (70°)	N. 70°W (290°)	S. 70°E (110°)	S. 70°W (250°)	0.3	0.4	0.4	0.5	0.6	0.7	0.8	0.9	1.0	1.1	1.1	1.2	1.3
N. 80°E (80°)	N. 80°W (280°)	S. 80°E (100°)	S. 80°W (260°)	0.1	0.2	0.2	0.3	0.3	0.4	0.4	0.5	0.5	0.6	0.6	0.7	0.8
E (90°)	W (270°)	E (90°)	W (270°)	—	—	—	—	—	—	—	—	—	—	—	—	—

LATITUDE 70° (NORTH OR SOUTH).

δ Westerly. Course.		δ Easterly. Course.		Speed in Knots.												
				6	8	10	12	14	16	18	20	22	24	26	28	30
N. (0°)	N. (0°)	S (180°)	S (180°)	1.1	1.5	1.9	2.2	2.6	3.0	3.4	3.7	4.1	4.5	4.8	5.2	5.6
N. 10°E (10°)	N. 10°W (350°)	S. 10°E (170°)	S. 10°W (190°)	1.1	1.5	1.8	2.2	2.6	2.9	3.3	3.7	4.0	4.4	4.7	5.1	5.5
N. 20°E (20°)	N. 20°W (340°)	S. 20°E (160°)	S. 20°W (200°)	1.0	1.4	1.7	2.1	2.4	2.8	3.2	3.5	3.8	4.2	4.5	4.8	5.2
N. 30°E (30°)	N. 30°W (330°)	S. 30°E (150°)	S. 30°W (210°)	0.9	1.3	1.6	2.0	2.3	2.6	2.9	3.4	3.6	3.9	4.1	4.5	4.8
N. 40°E (40°)	N. 40°W (320°)	S. 40°E (140°)	S. 40°W (220°)	0.8	1.2	1.4	1.7	2.0	2.3	2.6	2.9	3.1	3.5	3.5	4.0	4.3
N. 50°E (50°)	N. 50°W (310°)	S. 50°E (130°)	S. 50°W (230°)	0.7	1.0	1.2	1.5	1.7	1.9	2.2	2.4	2.6	2.9	2.9	3.3	3.6
N. 60°E (60°)	N. 60°W (300°)	S. 60°E (120°)	S. 60°W (240°)	0.5	0.7	0.9	1.2	1.3	1.5	1.7	1.9	2.0	2.3	2.4	2.5	2.8
N. 70°E (70°)	N. 70°W (290°)	S. 70°E (110°)	S. 70°W (250°)	0.4	0.5	0.6	0.8	0.9	1.0	1.2	1.3	1.4	1.6	1.6	1.7	2.0
N. 80°E (80°)	N. 80°W (280°)	S. 80°E (100°)	S. 80°W (260°)	0.2	0.3	0.4	0.4	0.5	0.5	0.6	0.6	0.8	0.8	0.8	0.9	1.0
E (90°)	W (270°)	E (90°)	W (270°)	—	—	—	—	—	—	—	—	—	—	—	—	—

An apparent disadvantage of the Gyro Compass often urged against it, is the length of time necessary for it to take up a position definitely. If it is started up with the **north** and **south** poles on the card, a long way from the meridian, then **two** hours must elapse before it can be used for steering, and **three** hours before really accurate observations can be taken with it (see curve of damping, page 32).

It must be remembered that when a ship is going to sea, this fact is known some time in advance for the purpose of raising steam, etc. ; so that there is plenty of time to get the Gyro Compass going ; further, no harm is done by leaving the Gyro Compass running when in harbour, the wear upon the bearings being infinitesimal ; no appreciable wear being found in instruments which have run for 4,000 hours.

The time necessary for the Compass to settle down with the north and south points on the meridian line can be very materially reduced if the Card be slowly turned somewhere near north and south, by hand ; and at the same time steadied in a horizontal position (seen on the level on the card), as soon as it is running at its full speed.

The Gyro Compass **can** never make a sudden movement, owing to its great resisting power (due to rotation), even when heavy shocks are experienced.

The resistance of the rapidly rotating mass of the Gyro makes the Gyro Compass independent of errors due to lag, so that the smallest departure of the ship from her correct course can be observed ; the Gyro Compass is entirely free from the lag which takes place in a Magnetic Compass, and therefore is far easier to steer a straight course with.

This can be more readily observed in the receivers when transmitting gear is used, as in this case no error due to lag exists at all. The central dial of the receiver being divided into spaces each representing 6′ of arc, the course can be determined ten times as accurately as heretofore. The receiver dial is illustrated at Figure 21, page 41, and at page 43 a further explanation of the central dial is given.

For commercial ships the employment of the transmitting mechanism and receivers enables a **shorter distance** to be steamed, and a consequent **increase of speed** due to the immediate correction of any yaw, with a minimum of helm.

Laying out a course on a chart is much simplified, as only one correction has to be taken into consideration. This is only a very unimportant one and can be neglected altogether when no special requirements as to accuracy are made.

If a Gyro Compass is used alongside a Magnetic Compass, an excellent opportunity presents itself for determining the deviation of the latter ; the indications of the Gyro Compass are free from deviation, so that if the ship is slowly swung completely round through the whole 32 points in the ordinary manner for adjusting compasses, simultaneous readings indicate the condition of affairs as regards the Magnetic Compass.

At the same time the correct position for the lubber point in the Gyro Compass can be accurately determined by taking the mean of the differences between the Magnetic Compass readings and the Gyro Compass readings all round the 32 points.

Transmission.

The general arrangement of the transmitter and receiver dials somewhat resembles an electrical engine-room telegraph, the signals are transmitted steadily, there is therefore no risk of the receivers getting out of step.

It is obvious that in a large ship the apparatus **cannot** be called upon to make a complete turn of 32 points in less than 6 minutes.

When under way, the small central Compass Card is constantly "on the move," on account of the ship's continual departure from an absolutely true course. This movement serves as an indication that everything is working. A continued movement in either direction shows that an alteration of course is taking place, see Figure 23, page 43.

The receivers all turn in the **same direction.** They can be installed in any convenient position; only the receiver which is intended for use with an azimuth mirror, or device for taking bearings, must be fixed horizontally in the centre line of the ship, with the lubber point fore and aft.

By suitably installing several receivers, it is possible to take bearings all round the horizon, clear of all obstructions, and several observers can take simultaneous bearings with certainty that their Compass Receivers are exactly in agreement.

The receivers are **independent** of position, and can, if necessary, be connected up by means of flexible cable. In the case of the Steering Compass, the dial can be inclined, or it can be fixed vertically.

Accuracy.

The accuracy of the Gyro Compass depends to some small extent on the position selected for it in the ship ; the most favourable position is, naturally, at the metacentric line of the ship. A position fairly free from vibration on a well-stiffened deck is desirable. If these general conditions exist, and no excessive fluctuations in pressure take place on the ship's lighting circuit, then an accuracy of 1° in either direction can be counted upon.

If it is inevitable that unfavourable conditions in these respects have to be put up with, then an error of 2° either way may have to be reckoned with. It is recommended that the navigating officer should determine for himself, by means of a series of observations after the Compass has been installed, what the accuracy of the Gyro Compass really is under the particular conditions.

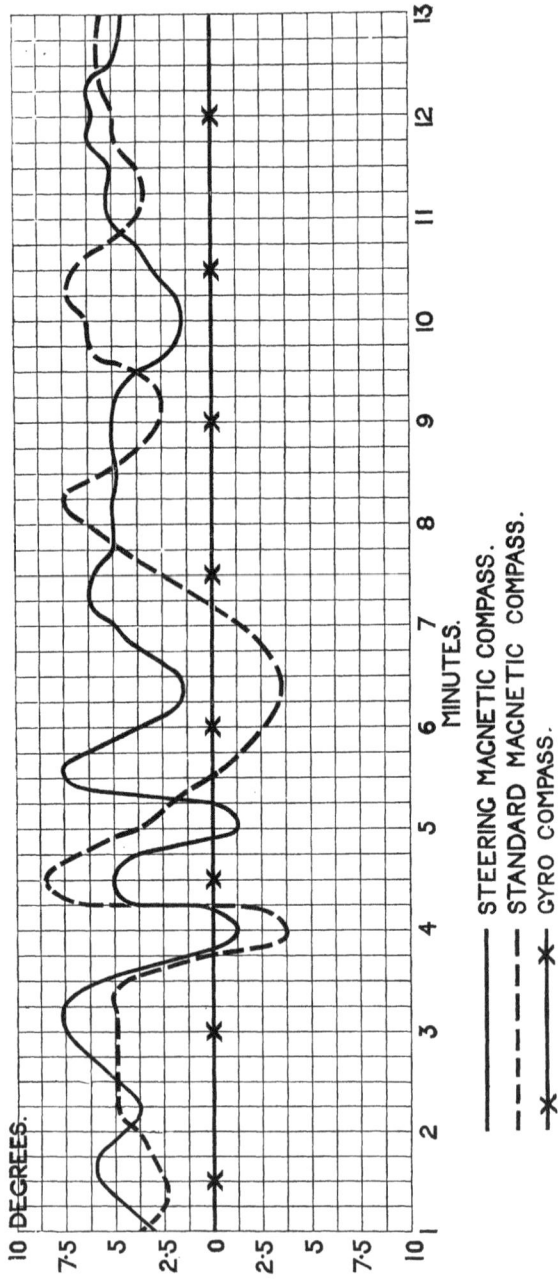

Fig. 39.

If especially unfavourable conditions exist for the use of the **single** instrument the employment of the **transmitter** and **receiver** system will prove of great assistance ; because then the Gyro itself can be placed low down in the ship in some position where the least vibration exists.

Figure 38, page 85, shows graphically the curves of oscillation of Magnetic Compasses from the effect of gun-fire— it can be seen from this that there is no effect from gun-fire on the Gyro Compass. A great many very exhaustive trials have been made in this direction during gun trials.

In conclusion it may be stated that criticism will be warmly welcomed by the makers, as they always have experiments in hand, and improvements are constantly being introduced as experience under sea-going conditions is available to guide them.

It would be of great interest to learn all details of the differences between Gyro Compass readings and Magnetic Compass readings under special conditions, such as passing a large iron ship, hoisting in steam boats, heating of steel work by sun's rays, and so forth.

Installation and Maintenance.

Instructions for using the Gyro Compass.

First of all the transformer must be connected up to the Gyro Compass before it is started up :—

By closing the three-pole switch at the left of the switch-board. It is of course immaterial whether this is turned up or down, as the two positions only deal with the two sets of fuses.

The voltmeter switch should be on one of the outside sets of studs.

The whole of the pressure regulating resistance in the rheostat at the right-hand lower part of the board should be cut out by turning the arm into the position at the extreme left, so as to give the maximum excitation to the three-phase generator.

The speed-regulating switch fixed at the left-hand lowest part of the board should be turned to the extreme right, so as to insert the whole of its resistance.

This refers to the rheostat for the motor portion of the motor generator.

Next the starting switch situated between the two rheostats should be turned up so as to start the motor running.

The current on all 3 ammeters will now be seen to rise gradually. The pressure-regulating switch is turned gradually over to the right, contact by contact, taking care that the current never exceeds ·8 ampere. This should be done very slowly, and while it is taking place it will be noticed that the armature of the motor generator revolves very slowly indeed, far too slow, in fact, for its speed to be shown on the speed indicator.

The whole process of starting up should take at least 20 minutes before the Gyro can be run safely up to its full speed of 20,000 revolutions per minute.

As soon as the pressure-regulating rheostat has been turned to the extreme right step by step, time should be allowed to elapse until the current falls to ·7 ampere. The fall of the current is, of course, due to the gradually increasing back electromotive force of the Gyro Motor as it gathers speed.

The speed regulating-resistance should be now turned round one contact towards the left, and at the same time the position of the pressure regulating resistance should be so adjusted that the current on no account exceeds 1 ampere, by turning the regulating switch one or two contacts to the left.

The speed should then be increased gradually by turning the speed-regulating rheostat very slowly to the left, and always so manipulating the pressure-regulating rheostat that the current does not exceed 1 ampere, this procedure being carried on until the speed indicator attached to the motor generator shows that this is running at 1,000 revolutions per minute.

The same manipulation should be carried on, but when the speed of the motor generator exceeds 1,000 revolutions per minute more current can be allowed to run to the Gyro Motor. Current can be allowed to run up to 1·4 amperes.

From time to time while this is being done it is advisable to change the voltmeter switch into each of its three positions, and observe the reading of the voltmeter to see that equal pressures exist across each phase.

On no account should the current be allowed at any time to exceed 1·5 amperes.

Correct Direction of Rotation of Gyro.

This should present no difficulty at all if the connections are properly made, and the scheme of colours adopted throughout the whole of the three-phase connections ensures this.

It can at once be seen whether the Gyro is rotating in the proper direction by observing the spirit level fixed on the top of the Compass Card.

The Compass Card will be seen to precess round, if the Gyro is started up with the north and south points on the card not coinciding with the meridian, as will almost invariably be the case, the precession at low speeds being fairly rapid.

The spirit level travels round with its **lowest point first,** that is to say, the little air bubble is at the back part of the spirit level in its rotation.

Should the Gyro be started up in the wrong direction the red and white wires should be changed round.

The Gyro should be run gradually up to speed as described above until the speed indicator attached to the motor generator indicates 2,500 revolutions per minute. When this point is reached the pressure-regulating rheostat should be manipulated so that the indication of the voltmeter is 120 volts across either of the three phases. The current under these conditions will be seen to be about 1·1 amperes, and this is the normal condition for running continuously.

While starting up, should a fuse give way, either in the continuous current circuit supplying current to the switch-board from the ships lighting circuit, or one of the fuses from the three-phase circuit between the motor generator and the Gyro Motor, then the following points must be observed:—

If the continuous current supply circuit fails, the automatic coil on the starting switch will cause this to be thrown out, and the motor generator would stop running. The Gyro, of course, will continue running for a very long time owing to its immense inertia.

The three-pole switch at the left of the board should then be placed in its open position.

The motor generator should be started up again and run up to speed by the speed regulator until the speed indicator on the motor generator indicates 2,500, and at the same time the pressure-regulating rheostat should be turned to the extreme right so that the minimum excitation is in existence.

The three-pole switch should then be closed and the speed-regulating rheostat should then be adjusted until the current on the ammeters drops back to about 1 ampere, which shows that the normal condition of affairs has been restored.

The above process should be gone through if for any reason it is desired to stop the motor generator purposely while the Gyro is still running, and later start up the motor generator and get it again into synchronism.

Should one of the fuses in the three-phase circuit fail, the whole system will go on running with two phases only connected for a quarter of an hour without any damage being done. The fact that only two phases are connected will be seen by the readings of the ammeters, two of the ammeters reading much in excess of their normal current, and the third falling back to zero. This, of course, shows which fuse has gone.

The Gyro Motor should not be allowed to run long with only two phases, as the windings become overheated.

The three-pole switch on the left of the switchboard should be at once changed over into the other position so as to bring the other three fuses into circuit, when the fuse which has failed can be replaced at leisure.

Point for attention while running.

The normal speed of the motor generator should be kept accurately at 2,500 r.p.m., from time to time the ammeters should be looked at to make sure that the current in all three phases is the same, and the speed indicator should be observed.

The motor generator requires the ordinary care given to any dynamo or motor, and it should run without sparking. If necessary the commutator should be cleaned with very fine glass paper (not emery). The brushes can be changed while the motor generator is running. The oil cups of the speed indicator should have a few drops of oil occasionally, and if the speed pointer vibrates, the coupling to the shaft should be examined at the first opportunity and renewed if necessary.

Stopping Down.

First place the three-pole switch at the left hand side of the board in the central or off position, and then touch the lever of the automatic portion of the starting switch so as to trip this off.

The pressure-regulating switch and the starting switch should be turned into the proper positions for starting up again, otherwise the interlocking gear between the speed regulator and the starting switch will prevent the starting switch being closed.

The Gyro will continue to run for a considerable time, and should be allowed to stop by itself.

Oiling the Gyro.

Two oil cups are provided on the Gyro which should be examined daily, and if the oil is discoloured or if the cups are half empty, they should be removed and refilled with the special oil supplied with the Gyro Compass. The oil should in any case be renewed once a week. Only the special oil should be used, as experience has shown this to be the best, and the very greatest care should be taken when removing, filling, and replacing the oil cups so that by no chance any grit or dirt gets in. The oil cups should be touched as little as possible, to avoid disturbing the Compass.

Failure of One Phase.

If trouble is due to a defect on one wire it is not possible to run at full speed with two phases, and the speed of the motor generator should be reduced to 1,600 to 1,800 r.p.m., care being taken that the current does not exceed 1·5 amperes. The Gyro Compass loses some of its accuracy, but can still be used as a steering compass. If the Gyro is stopped it cannot be started up again with two phases.

Reversible Motor in Binnacle.

When the transmission system is employed this motor runs whenever the ship alters her course, and on that account attention must be paid to the filling of the lubricator, which can be got at when the door of the binnacle is opened.

Receivers.

These do **not** require oiling.

The adjustment of the receivers can be done by first removing a screwed cover at one side of the case and then turning the small stem in the centre of the opening, by this means the indications of the receivers can be adjusted to correspond with those of the Master Compass.

Receiver Circuits.

At one point only in these are fuses, in a fuse box close to the binnacle.

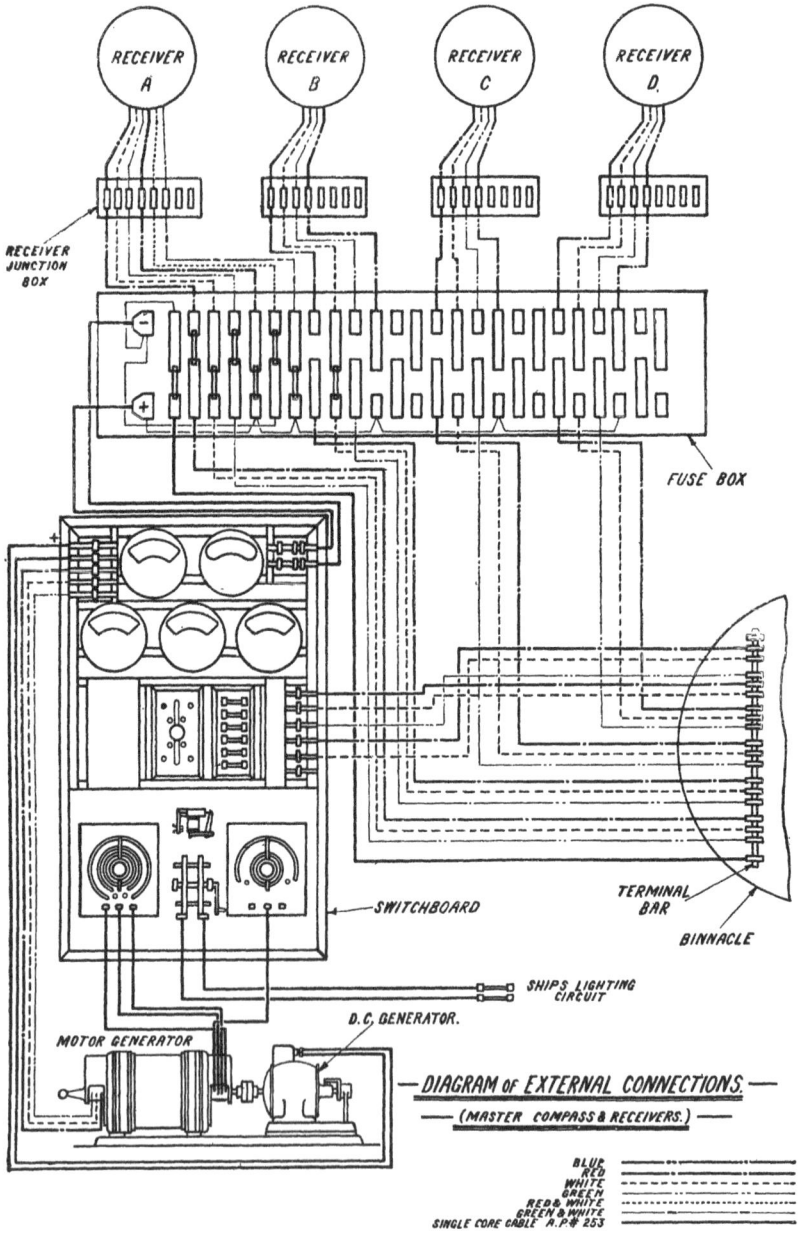

RECEIVER A RECEIVER B RECEIVER C RECEIVER D

RECEIVER JUNCTION BOX

FUSE BOX

TERMINAL BAR

BINNACLE

SWITCHBOARD

SHIPS LIGHTING CIRCUIT

D.C. GENERATOR.

MOTOR GENERATOR

DIAGRAM of EXTERNAL CONNECTIONS.

(MASTER COMPASS & RECEIVERS.)

BLUE
RED
WHITE
GREEN
RED & WHITE
GREEN & WHITE
SINGLE CORE CABLE A.P# 253

Fig. 40.

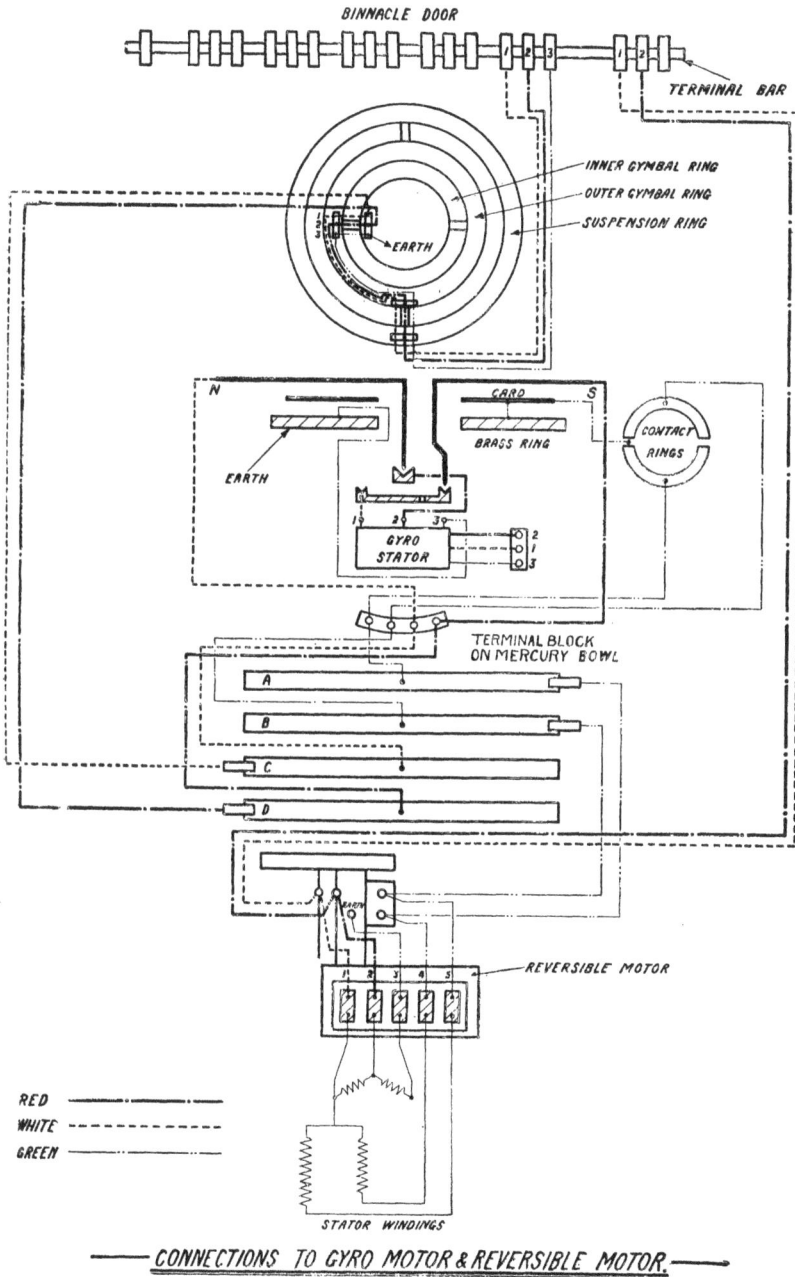

CONNECTIONS TO GYRO MOTOR & REVERSIBLE MOTOR.

Fig. 41.

H

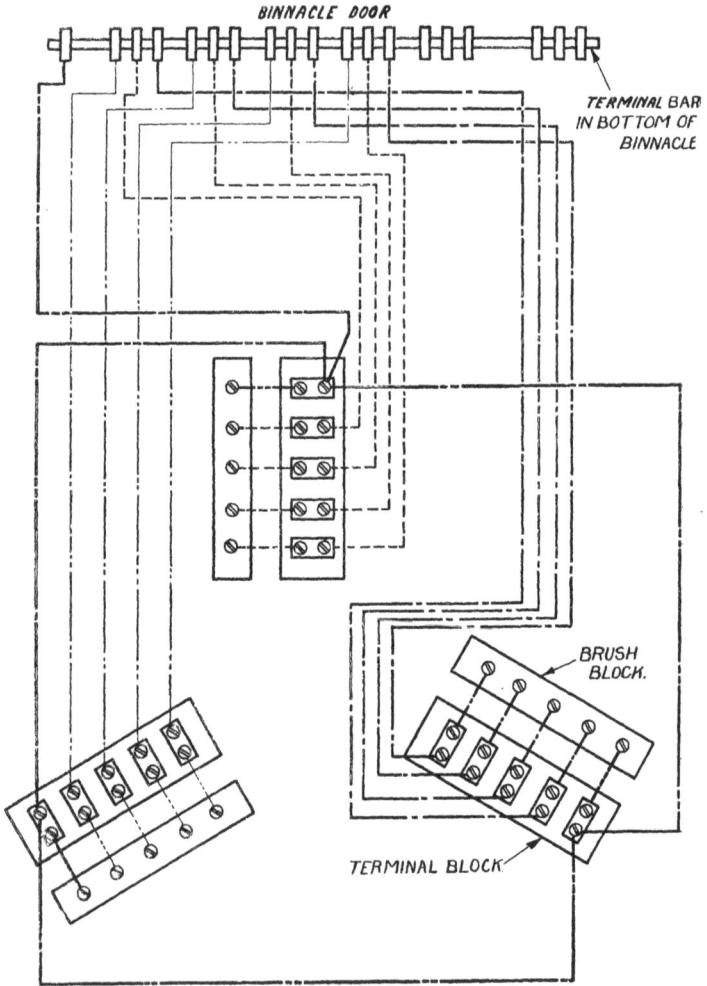

BINNACLE DOOR

TERMINAL BAR
IN BOTTOM OF
BINNACLE

BRUSH
BLOCK.

TERMINAL BLOCK

BLUE.
RED.
WHITE.
GREEN.

CONNECTIONS TO COMMUTATOR

Fig. 42.

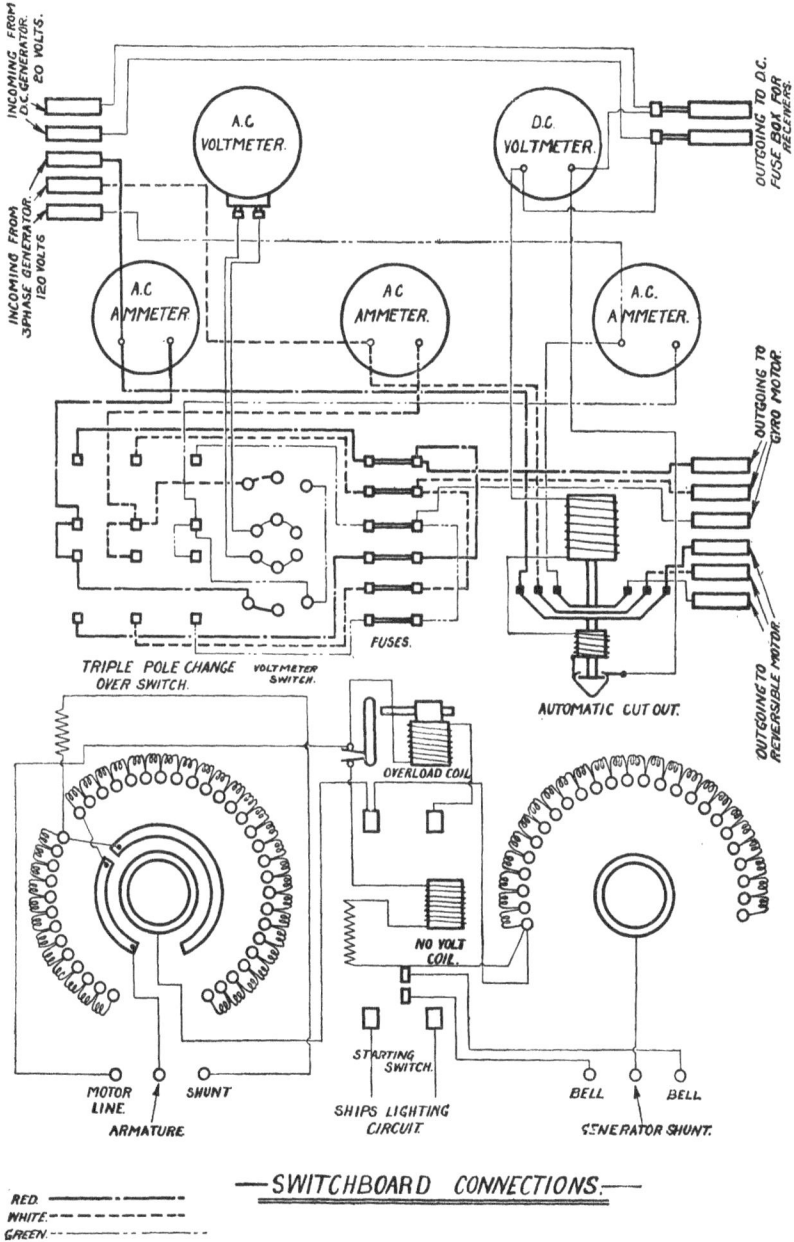

SWITCHBOARD CONNECTIONS.

Fig. 43.

108

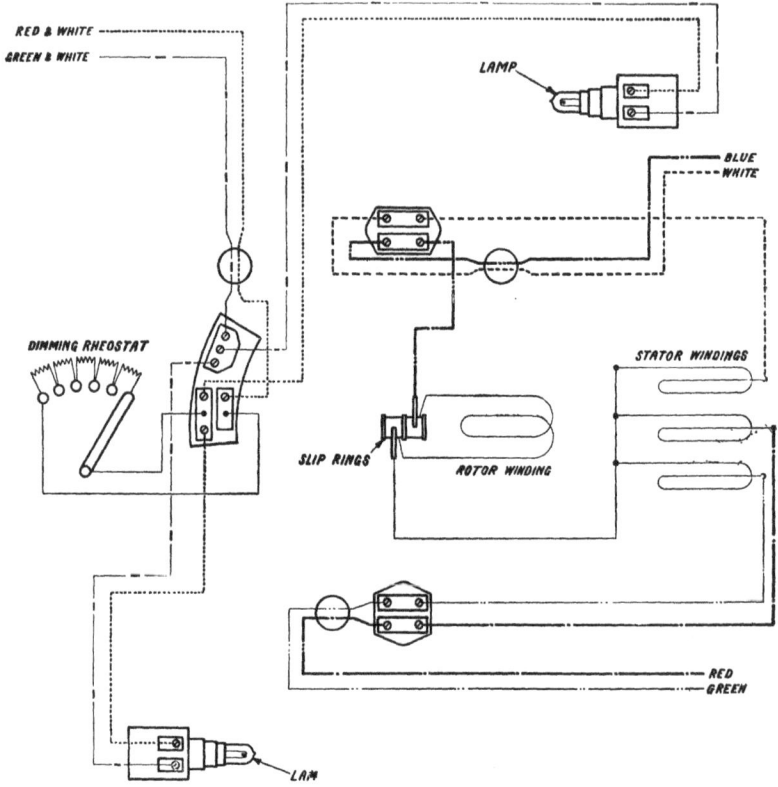

CONNECTIONS FOR ILLUMINATED RECEIVER.

Fig. 44.

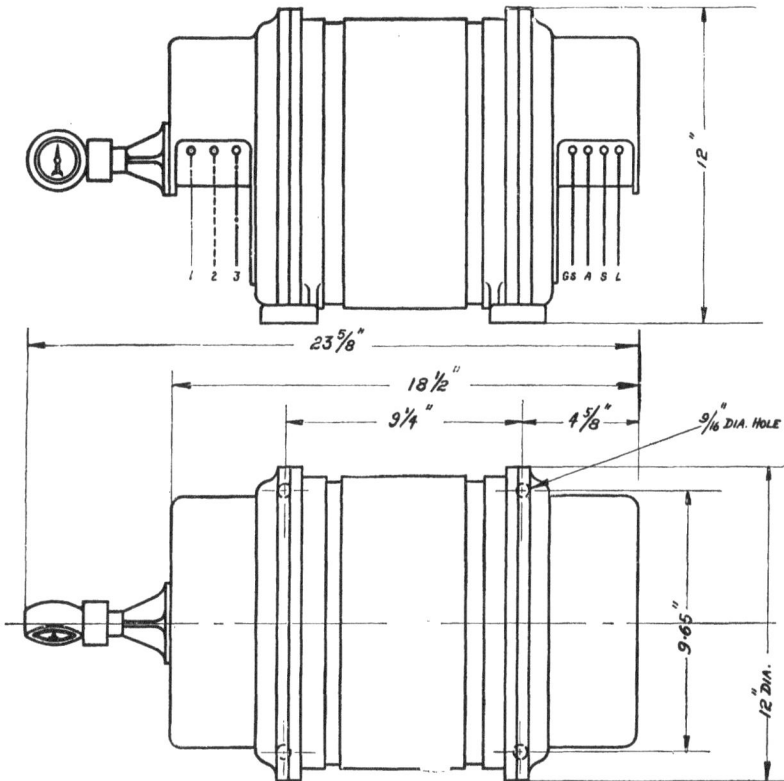

MOTOR GENERATOR.

GS = GENERATOR SHUNT
A = ARMATURE
S = SHUNT
L = MOTOR LINE

RED
WHITE
GREEN
SINGLE CORE CABLE A.F.#253

Fig. 45.

Index.

www.ingramcontent.com/pod-product-compliance
Lightning Source LLC
Chambersburg PA
CBHW031402180326
41458CB00043B/6584/J